T0243460

Praise for *Barking Up the Right Tree*

"Ian Dunbar is the godfather of positive dog training, and his decades of experience are distilled in this one *essential* book."

— **Larry Kay**, coauthor of *Training the Best Dog Ever* and *The Big Book of Tricks for the Best Dog Ever*

"Dr. Ian Dunbar is quite simply Britain's greatest export. He showed the dog world how to play nicely, and this book is every bit as rewarding as his training methods. Dunbar's gentle wit makes his enormous wisdom jump off the page straight into your DNA. I can't believe he actually finally got around to writing it all down. This is a timeless bible for anyone who wants to be their dog's hero, an instant classic. He's always been my guru; now he can be yours, too."

— **Beverley Cuddy**, editor of *Dogs Today* magazine for thirty-three years

"Dr. Dunbar's latest book is a testament to his deep-rooted expertise and pioneering contributions to the field of dog training. His emphasis on lure-reward training techniques showcases a commitment to positive methods that prioritize the dog's emotional well-being. This work is a valuable addition to any dog lover's library, offering a blend of time-tested wisdom and innovative approaches."

— **Zak George**, bestselling author of *Zak George's Dog Training Revolution*

"A wholly enjoyable and informative book that will surely have an impact on dog lovers and trainers alike."

— **Julie Hecht**, canine behavioral researcher and science writer

"If you want a great relationship with your dog that is free from stress and heartache, read this book. Through stories and clear examples drawn from decades of indelible work in the field, Ian Dunbar paints a picture of dog training that is as easy and fun as it is effective and grounded. He has a magical way of lifting dog training out of the sterility of scientific discourse and breathing it to life in the real world. Using his unique wit, he somehow dissolves the polarizing tensions of the dog world and delivers tangible words of wisdom we can all live by."

— **Amanda Gagnon**, anthrozoologist, dog behavior consultant, and dog trainer in New York City

"*Barking Up the Right Tree* compiles Ian Dunbar's lifetime of research, education, and teaching into an educational and entertaining blueprint for trainers and pet owners everywhere. It's a must-read that shines a bright light on the future of positive training for dogs and the people who love them. Woof!!"

— **Sue Pearson, MA**, SPOT & CO. Dog Training

"Dr. Ian Dunbar has authored some of the most important and useful dog training books using positive reinforcement and a technique called luring, which he details in this book. Dunbar explains complex learning theory in simple terms and how to motivate dogs with force-free methods. Simply put, Dunbar is a legend in dog training — and all words he speaks are worth paying attention to, including an entire chapter expressing his concerns regarding modern dog breeding."

— **Steve Dale, CABC**, host of the nationally syndicated *Steve Dale's Pet World* (also heard on WGN Radio, Chicago; SteveDalePetWorld.com) and coeditor of *Decoding Your Dog*

"In this wise and funny book, Ian Dunbar unpacks how he teaches dogs ESL. But what he's also doing is teaching people DSL — dog as a second language. Dunbar has been speaking it for decades, and it's a pleasure to hear him translate dogs for us again."

— **Alexandra Horowitz**, author of the *New York Times* bestseller *Inside of a Dog: What Dogs See, Smell, and Know* and *The Year of the Puppy: How Dogs Become Themselves*

"World-renowned dog trainer Dr. Ian Dunbar does it once again. *Barking Up the Right Tree* is an encyclopedic, evidence-based, and easy-to-read discussion of positive dog training. His approach is the only way to teach dogs what we would like them to do and not do, learn and unlearn, while respecting them for their rich and deep cognitive and emotional lives and honoring the well-established fact that they are fully feeling, sentient beings who deserve nothing less than the very best and most respected lives. I am sure *Barking Up the Right Tree* will be celebrated as the bible of dog training as more and more people come to recognize that there are no humane alternatives to Dunbar's approach."

— **Marc Bekoff, PhD**, coauthor of *Unleashing Your Dog: A Field Guide to Giving Your Canine Companion the Best Life Possible* (with Jessica Pierce) and author of *Dogs Demystified: An A-to-Z Guide to All Things Canine*

"Ian Dunbar has long been known as one of the dog world's most influential innovators. He has not only taught thousands of professional dog trainers and founded the largest dog training organization in the world, he was also instrumental in establishing the importance of off-leash puppy socialization and the positive approach to dog training. His influence is wide-ranging and frankly incalculable. And now, in *Barking Up the Right Tree*, he explains — using the wit, warmth and wisdom he's known for — why the lure-reward training method is the most effective and humane way to train our dogs. He teaches us how to think like a dog and also spells out the easy steps that will help our dogs understand our language, which, as he says, 'is the quintessence of lure-reward training.' Ian Dunbar's legacy is already profound, and this engaging and informative book is yet another reason to be thankful for all he has done for dogs and dog lovers everywhere."

— **Claudia Kawczynska**, founder of *The Bark* magazine and former editor in chief of *The Bark* and TheBark.com

"What else is there to do but howl with gratitude and delight that Dr. Ian Dunbar has written this outstanding book?! No one better understands what motivates dogs, and the ways to maximize desired behaviors to yield a human-canine relationship that truly works both ways, than the inimitable Dr. Dunbar. For over fifty years he has joyfully and wholeheartedly been at the forefront of positive dog training, watching what works and doesn't work with dogs and people around the world. In a Google search for 'best dog trainers in the world,' it's no wonder his name is first on the list. Want to know what to do about barking, whining, digging, pulling, inappropriate elimination, separation anxiety, and so many more behaviors that can derail our relationships with our dogs? Want to understand how to really be your dog's best friend? Then read *Barking Up the Right Tree*. Your relationships with not just your dog(s) but your friends, family, co-workers, other animals, and the planet will be forever changed — in a practical and oh-so-positive way. Thank you once again, Ian Dunbar."

— **Dominique DeVito**, former publisher at Howell Book House, TFH Publications, and IDG Books and senior editor of the *AKC Gazette*

"Dr. Ian Dunbar took dog training from the dark ages, transforming the approach of many that followed from punitive to positive. His new book, *Barking Up the Right Tree*, draws on five decades of experience to offer specific, effective, and proven dog-centered techniques for a calm, well-behaved, happy pup that is a joy to be around. A must-read for dog owners wanting to improve their relationships with their dogs."

— **Jennifer Nosek**, editor of *Modern Dog*

"This book is a powerful, convincing love song to happy, reward-based training. Any trainer looking for help convincing a client not to use aversive methods just needs to memorize Dunbar's science-filled, client-tested, master-class chapter on punishment. I had wondered how a man had landed in the jerk-and-correct dog industry of fifty years ago and very publicly concluded 'bunk.' It's fascinating to read that it was Ian Dunbar's childhood on his grandfather's farm that inoculated him against aversive training groupthink. He learned as a five-year-old that you can't leash-and-force a heifer, but you sure can use a lure, and become friends. Thanks, Grandpa, for helping your grandson learn lessons he'd carry out into the world to help dogs. Dunbar's fifty years of working with actual clients and their dogs is evident throughout this book, as he makes a convincing case for the simplest, most effective methods for producing a joyful human-canine relationship."

— **Kathy Callahan, CPDT-KA**, author of *101 Rescue Puppies* and *Welcoming Your Puppy from Planet Dog*

BARKING UP THE RIGHT TREE

Also by Dr. Ian Dunbar

Before and After Getting Your Puppy:
The Positive Approach to Raising a Happy, Healthy & Well-Behaved Dog

Doctor Dunbar's Good Little Dog Book

How to Teach a New Dog Old Tricks:
The SIRIUS Puppy Training Manual

BARKING UP THE RIGHT TREE

THE SCIENCE AND PRACTICE OF POSITIVE DOG TRAINING

DR. IAN DUNBAR

New World Library
Novato, California

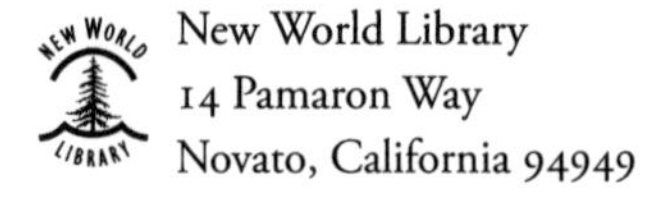
New World Library
14 Pamaron Way
Novato, California 94949

Text design by Tona Pearce Myers

Library of Congress Cataloging-in-Publication Data

Names: Dunbar, Ian, date, author.
Title: Barking up the right tree : the science and practice of positive dog training / Dr. Ian Dunbar.
Description: Novato, California : New World Library, [2023] | Includes bibliographical references and index. | Summary: "Based on decades of dog behavioral science, *Barking Up the Right Tree* is a guide for dog owners looking to improve their relationship with their canine companions and ensure that their dogs become calm, confident, well behaved, and happy"-- Provided by publisher.
Identifiers: LCCN 2023033892 (print) | LCCN 2023033893 (ebook) | ISBN 9781608687718 (hardback) | ISBN 9781608687725 (epub)
Subjects: LCSH: Dogs--Training. | Dogs--Behavior. | Dogs--Psychology.
Classification: LCC SF431 .D7697 2023 (print) | LCC SF431 (ebook) | DDC 636.7/0887--dc23/eng/20230912
LC record available at https://lccn.loc.gov/2023033892
LC ebook record available at https://lccn.loc.gov/2023033893

First printing, December 2023
ISBN 978-1-60868-771-8
Ebook ISBN 978-1-60868-772-5
Printed in the United States on 30% postconsumer-waste recycled paper

New World Library is proud to be a Gold Certified Environmentally Responsible Publisher. Publisher certification awarded by Green Press Initiative.

10 9 8 7 6 5 4 3 2 1

The best time to plant a tree is twenty years ago;
the second-best time is today.

— Attributed to a Chinese/African/Greek proverb

Contents

Introduction

Barking Up the Right Tree

Barking Up the Right Tree is very different from any book I have written before. Most of my books and videos have been about dog social and sexual behavior, or comprehensive guides and how-to, step-by-step instructions for new puppy owners, such as *Before & After Getting Your Puppy*.

This book focuses on describing and teaching lure-reward training. This involves (1) quickly and easily teaching a dog ESL (English as a second language), that is, *the meaning* of the words we use; (2) *testing comprehension* of verbal instructions; (3) using much more powerful *life rewards*, such as walks and dog-dog play, and *games* to create a self-motivated dog; and (4) using various effective, *nonaversive* techniques for resolving misbehavior and lack of compliance. Also, I provide guided instruction for the prevention and resolution of most of the more common behavior, temperament, and training problems that people encounter.

The premise of this book is simple: It's high time we celebrate the glory of dog-friendly, people-friendly, quick-and-easy, fun-and-games, off-leash, results-based, *lure-reward dog training*. I have developed and refined these techniques over the past fifty years. Lure-reward training is the *easiest*, *quickest*, and *most effective* technique to teach your dog the meaning of your words, and so open communication channels to considerably facilitate teaching basic manners, preventing or resolving behavior and training problems, and progressively molding your dog's temperament and personality.

In addition to teaching dogs to Come, Sit, and walk calmly on-leash, dog training needs to give dogs and puppies the *gift of confidence* around people and other dogs, the skills to handle being alone, comprehension of our language for clear instruction and ongoing guidance when they go off track, and through games, praise, and a wide range of *life rewards*, the motivation to *want* to do what we ask.

At root, my goal is always to consider both the dog's point of view and the person's point of view and to develop techniques that can work for anyone and everyone: for dogs of any age and breed and for all types of people and their families, especially including children.

To avoid and solve unwanted behavioral issues, people must understand dog behavior, learn to "read" their dog's feelings, especially anxiety or stress related to people and other dogs, and cater to their dog's needs. Essentially, reward-based training gives dogs what they need and want in ways that inspire them to do what we need and want.

The essence of *Barking Up the Right Tree* is to explain lure-reward training so that you understand not only how to communicate with, motivate, and change the behavior and temperament of your dog, but also why these methods work so well.

If This Is Wrong, What's Right?

"My dog won't listen to me!"

"He knows it's wrong!"

"How do I stop my dog from housesoiling, chewing, and barking?"

"How can I punish my dog when he's off-leash?"

"How can I punish my dog when she misbehaves when left at home alone?"

Punishment has often been sold as a quick fix for problems like these, but in fact, most of the time, punishment is neither quick nor a fix. On-leash, punishment-based training is often time-consuming and relatively ineffective. The focus is usually on disobedience and behavior problems and on inhibiting and eliminating a dog's natural behavior, rather than teaching dogs *how we would like them to act.*

Early on in my career as a dog trainer, I tried to devise ways to modify dog behavior as quickly and easily as possible while remaining responsive to the dog's desires and delights. At that time, my wife Mimi and I were struggling with a chewing habit of our first dog, Omaha, an Alaskan Malamute. Over his first few months at home, slowly but progressively, Omaha had reduced the size of our front hall carpet from a neat eight-by-ten-foot rectangle into a raggedy-edged, eighteen-by-twenty-four-inch doormat.

Trained as a developmental psychologist with an emphasis on cognition, Mimi's approach to training (that is, teaching) our puppy (and later, our son, Jamie) always championed "needs and feelings," whereas I tended to emphasize observing, quantifying, and then changing behavior and temperament. During our many puppy-raising discussions, Mimi asked me one question that changed my entire outlook on dog training forever — in addition to inspiring the resolution to Omaha's carpet-trimming "problem." Mimi's questions were always demurely asked, infuriatingly logical, and dagger-to-the-heart revealing.

She asked me, "Have you taught Omaha what you would like him to chew?"

I combined our two approaches and now, whenever someone comes to me with a dog problem, this is one of the first questions I ask. If, by the person's definition, a dog is doing something "wrong," then what is right? Define and teach that till it becomes second nature. The question also applies to evaluating the effectiveness of any dog training technique: If one approach isn't resulting in improvement, then try another approach. Don't keep doing the same thing over and over if it's not working.

How did I stop Omaha from chewing our carpet? I encouraged him to chew something else. This was before the days of chewtoys, so I improvised and resolved two problems in one fell swoop. Omaha was also excavating my garden, and I'm a gardener — *Arrrghh!* So I bought an entire meaty cow's femur from the Magical Meat Boutique, and one night after Omaha had bedded down, I went outside and buried the bone in a small, three-by-five-foot area under the rear deck stairs. The next morning, we went outside, and after I rewarded Omaha for eliminating in his toilet area, he set about sniffing the garden perimeter, and then... his nose elevated and started twitching. He caught the buried bone's spoor and took barely a minute to find its location.

And that was pretty much it. No more chewing on furniture and fittings; all chewing became directed to bones that he found in what I dubbed his "digging pit." Every night, I would bury a smaller bone freshly stuffed with mushed kibble and a variety of treats for him to find in the morning. This also solved the issue of digging in the lawn and flower beds. Consider Omaha's point of view: *Wow! There's hardly anything of doggy interest buried in this entire garden, but this digging pit is loaded with buried treasure!* This is why, in 1848, a considerable portion of the US population ventured west to California — due to a single find of gold in Sutter Creek, but next to nothing in New Jersey.

I still have the gnawed and bleached remnant of Omaha's original cow femur. It sits on top of a bookcase where I can see it every evening.

Omaha's education, and further long discussions with Mimi, taught me that "a dog's a dog," and that for training to be successful, we *must* cater to their specific needs and feelings. Rather than attempt to "take the dog out of the dog," my focus increasingly emphasized teaching dogs how to appropriately express their natural dogginess in ways that don't upset owners. For example, teaching and verbally cuing dogs *where* to eliminate, *what* to chew, *when* to bark, *when* to shush, and *how* not to upset other dogs that may not have the same degree of social savvy and confidence. A well-trained dog can do lots to calm other dogs that are fearful and reactive.

What Is Lure-Reward Training?

Dog training is almost entirely about *communication*, specifically teaching dogs *our language*, that is, ESL. Lure-reward training is by far the quickest and easiest technique for teaching dogs the meaning of our verbal instructions and for regularly testing comprehension as proof of training.

Initially, food lures are used to teach the meaning of handsignals, and then food lures are phased out and handsignals are used to teach the meaning of verbal instructions. Dogs learn handsignals quickly because they are a language dogs understand — body language. Once dogs *understand what we say*, dog training transcends to a different level. As our dog acquires an ever-expanding vocabulary, teaching basic manners becomes even quicker and easier.

The central tenet of any educational endeavor is to, first, clearly communicate what we would like someone to do. With our dogs, that means whatever is contextually *appropriate* from a human viewpoint, and then we reward them for doing it. Luring stacks the deck so the dog is more likely to get it "right" from the outset, and then the dog is frequently rewarded for getting it right, which is the prime directive of *all* reward-based training techniques.

As a dog's vocabulary grows, this changes the playing field when it comes to misbehavior and lack of compliance. Now, we may simply instruct our dog exactly how we would like them to act. It is an amazing relief and a true delight to learn how to *effectively* prevent or terminate misbehavior and noncompliance using our voice. In fact, voicing only a single word can communicate three vital pieces of information to the dog: (1) Stop what you're doing, (2) do this instead, and (3) the degree of danger for noncompliance.

For example, if your dog is about to pee in the house, say, with some urgency, "Outside" or "Toilet." If your dog is barking, instruct, "Shush." If your dog is practicing agility in the living room, say, "Bed." And if your dog is chasing the cat or your children, or about to dash out the front door or jump up on someone, simply instruct, "Sit." A simple Sit is often the solution to so many problems.

Rather than waiting for utterly predictable bad habits and behavior problems to rear their ugly heads, and then attempting to correct them *after* the fact, lure-reward training takes the opposite approach — preventing predictable problems by teaching desirable behavior from the outset and establishing and internalizing good habits. Then, as these good habits are frequently reinforced, they increase in frequency and naturally crowd out unwanted behaviors. Lure-reward training rests on several key elements: early socialization, clear communication, off-leash training, and the use of life rewards, not simply food treats.

- **Early socialization:** Ideally, prior to eight weeks of age, puppies should grow up *indoors*, in an *enriched environment*, and *meet lots and lots of people* safely at home. The consequences of insufficient socialization with people during the first three months of life become apparent when dogs reach five to eight months of age. Fear, anxiety, and reactivity toward scary stimuli and situations, especially unfamiliar people and dogs, are all *adolescent-onset* behaviors that destroy manners and make life a misery for dogs and owners alike. However, it's never too late to socialize an older dog, should they lack confidence and life skills. However old your dog is when they join your family, start socializing right away.
- **Clear communication:** We need to *bring back our voice to training*. From the outset, we must teach the *meaning of our instructions* and then *test that dogs understand* them. We need to teach dogs ESL (or whatever language you prefer). Clear verbal instruction is essential for cuing basic manners and for providing guidance when dogs err. Then let's use our words to praise our dogs and celebrate with them when they do a good job.
- **Off-leash training:** First train your puppy or newly adopted adult dog *off-leash at home*, both indoors and outdoors, and only then attach the leash to your *trained* puppy for their first walk. Otherwise, your untrained dog will likely learn to pull on-leash during their very first walk. For inveterate pullers, words of guidance help considerably; for example, to instruct a dog to speed up when lagging, say, "Hustle," and to instruct a dog to slow down when forging or pulling, say, "Steady."
- **Life rewards:** To begin, lure-reward training uses food *lures* to teach dogs *what* we would like them to do, and it uses food *rewards* to motivate them to *want* to comply. However, food

> lures are phased out entirely as soon as dogs learn handsignals, and food rewards are largely replaced by far more powerful *life rewards*. This is done by integrating numerous, very short training interludes into walks, sniffs, play with other dogs, and interactive games with us. Additionally, we creatively "power up" praise as perhaps the most powerful *secondary* reinforcement on the planet.

Other behaviors and activities that dogs really enjoy are also used as rewards — behaviors that most people consider "problems." However, by putting "problem" behaviors *on cue*, we can teach dogs *when* it is OK to bark, hug, or let off steam (and when it is not). By using cued behaviors as rewards, the problematic notion of the undesired expression of a dog's normal, natural, *and necessary* behaviors becomes history, since we know how to turn them on ... and how to turn them *off*. All in all, life rewards are so much more powerful and effective than a mere food treat.

The key for a successful relationship is learning how to fine-tune and mold our dog's temperament, reliably cue desirable behavior, modify behavior in terms of appropriateness (when, where, what, how much, and how long) and quality (reliability, promptness, duration, precision, pizzazz, and panache), and motivate dogs in such a way that they are only too happy to oblige.

Eventually, external rewards become unnecessary because the dog becomes internally reinforced and self-motivated, and simply doing what we've asked becomes more than sufficient reward. This is the tango of training — an exquisite, interactive, unique choreography of life that dogs enjoy with their people, and people enjoy with their dogs. In a sense, the same verve that dogs have for tracking, running, chasing, hunting, herding, and pulling fuels the dog's joy for training, that is, interacting with and being part of a trusting, thoroughly enjoyable, totally encompassing relationship with us. This is when we realize that dogs are so much more than their genetic heritage. Each dog is an individual — a one-off — *our dog*. And this is when we fall in love.

So Old It's New Again

To tell you the truth, these techniques are not new. They are so old they are *new again*. My great-grandfather won a straight-line ploughing contest with no reins or whip; he guided his horse with verbal cues and provided rich verbal feedback. My grampa and dad used these techniques to train their gundogs, farm cats, and livestock. When I was five, I had to spend an extended period staying at the farm (my dad fell off the roof and fractured his skull), and during that time, I mastered group recalls of fourteen cats, eighty cows, and two hundred chickens, and I started applying my newfound skills to hamsters, rabbits, guinea pigs, budgerigars, and tortoises. I spent most of my days roaming the fields with three dogs off-leash — my dad's springer, my gramp's Labrador, and my uncle's Jack Russell, and they all stayed close. I was lucky to grow up on a farm and learn from watching my father, grandfather, and their farmworkers, who still used "the natural way" of raising and training domestic animals — off-leash, using food lures to teach clear instructions, and using oodles of praise, food rewards, and life rewards to reinforce.

All reward-based techniques make training more enjoyable for you and your dog, and once you've taught your dog to eliminate on cue in their toilet area; to enjoy settling down quietly and calmly while staying busy with chewtoys; to enjoy their new chewtoy hobby whenever left at home alone; to thoroughly enjoy the presence of family, friends, visitors, and strangers; to actively enjoy, or confidently tolerate, meeting and greeting other dogs in parks and on walks; and to otherwise remain calm and chill around the house and in the car — *what's left to get on your dog's case about?*

Motivating People Is Part of Dog Training

Owners, too, want training to be a natural process that is enjoyable and that is quick and easy to master. They want to have fun and feel proud and happy with their efforts and their dog's performance. Owners want to create and relish a loving relationship and to talk to their dog, whether giving clear instruction and guidance, praising their dog for a good job well done, offering guidance when they err, or just nattering about daily events.

Successful training gives people a sense of satisfaction and achievement, and so they feel much happier about their dog — a sense of joy — which of course causes their dog to feel happier. Reward-based training changes people and changes dogs; feel-good smiles become as common as tail wags.

Why are ease and speed of training so important? Well, if dog training techniques are too complicated, requiring a masters in psychology or exacting leash dexterity, they will be well beyond the means and skillset of most owners. If dog training techniques are excessively time-consuming, requiring the patience of Job, many people might not devote sufficient time. A dog trainer might retort, "If you don't have the time, then you shouldn't have a dog." Of course, once people have a dog, that's all water under the bridge. My approach has always been, "Well, you have a lovely dog. Let me teach you the easiest and quickest way to train that even allows for you to be inconsistent much of the time."

Many professional dog trainers tend to underestimate their own experience and expertise. They fail to realize that what is so easy, straightforward, and second nature to them is often confusing, challenging, time-consuming, and unnecessarily complicated for everyday people who just want a dog that minds. Dogs are my passion, and I could talk about their behavior and education for days on end (and often have done), but I recognize that most dog owners want solutions *today! Right now!* Moreover, they want techniques that are easy to learn, quick to administer, highly effective, and can be incorporated within their current routines and lifestyle. I have always designed training techniques with the dog's whole family in mind.

Looking back over my career with dogs, I realize that *it's all about people*. Absolutely, certainly, there is no other variable that has a greater effect on a dog's behavior, temperament, and training than the people in charge of the dog's education. The age, sex, breed, and breeding of the dog are all very, very secondary. For me, people training is the key to successful dog training. I've learned how people learn and discovered what accelerates their education. Merely "telling someone what to do" is far from sufficient, whether that someone is human or canine.

Any effective dog training technique also needs to motivate *people* to want to train their dogs. One of the best ways I've found to motivate owners is by teaching tricks and playing games. Nothing accelerates training more. Really, there is little difference between an adult gruffly commanding, "Down!" and a child pointing their finger and saying, "Bang!" Both dogs lie down, albeit one prone and the other supine. Teaching tricks and playing games turn a perceived chore into an owner's favorite activity with their dog.

Playing games has an equally important yet hidden agenda. In addition to being serious fun, playing games offers one of the most enjoyable ways to quantify a dog's performance, reliability, and ongoing improvement. Games demonstrate not only that training is working but how quickly it is working. Quantification of performance is not necessarily done to compare dogs (or owners) as they compete with one another. Rather, owners and dogs compete with themselves to establish personal bests and then ... surpass them.

Watching your puppy learn is simply beyond magical. There is nothing more fulfilling than seeing a rambunctious, rumbustious, bumbling puppy instantly transform into a statue — sitting with head cocked to one side and with riveted focus — following a single verbal request: "Hugo, Sit." Then, an instant later, to reactivate with abandon when told, "Free dog." Miss out on this, and you're missing out on one whole sack load of joy.

What Is the Right Tree?

I'm a behaviorist at heart and so my approach to dog training has always been based on facts rather than opinions or value judgments. At every step of the process, I ask three very simple but revealing questions: Did it work? How well did it work? And how long did it take?

Training is about causing observable, quantifiable behavior change, and so I routinely test comprehension of verbal cues and response reliability to provide *proof of training* and proof of the *speed of training.* Lure-reward training is not the only reward-based technique, and I include a comparison of a variety of reward-based techniques, evaluating their pros and cons in terms of their purpose, ease, speed, effectiveness, and enjoyment. All have their strengths and weaknesses. However, *lure-reward training* is unrivaled as the quickest means for "putting behaviors on cue," that is, for teaching ESL.

In chapter 14, I also do the same thing with punishment techniques. The results are gobsmacking. First, most aversive punishments *don't work,* since they seldom reduce or eliminate undesired behaviors. This is evidenced by their repeated and continued use. If undesirable behaviors were reduced, then the need for and use of aversive corrections would also be reduced until they were eventually eliminated. This is not what we see in dog classes or when people walk their dogs on-leash. While there are several reasons for this, the main reasons aversive corrections are so often *ineffective* are poor timing, inconsistency, and a lack of instructiveness.

Second, even when punishment is effective for reducing and eliminating undesirable behaviors, this is *insufficient* — the job is only half-done. Surely when a dog misbehaves, the goal is to get the dog back on track. Inhibiting a dog's natural behavior does *not* teach a dog what precisely we want them to do. On the other hand, lure-reward training literally shines for delivering clear verbal instruction and heavy-duty reinforcement.

Third, in a head-to-head competition for resolving misbehavior and noncompliance, lure-reward training and verbal instruction are considerably *more effective, easier, and quicker* than physical and aversive techniques. Lure-reward training wins paws down.

Lure-reward training is the best way to teach verbal instructions, which is what makes it "the right tree," and verbal instruction is the most effective means to reduce and eliminate undesirable behavior — which also eliminates the need for reprimands or aversive techniques.

This is why I often say that the use of aversive techniques is mostly *ineffective* but *always insufficient, unnecessary,* and therefore, unwarranted, especially given the numerous effective, nonaversive (reward-based) techniques at our disposal.

There is no doubt that the *jewel in the crown* of this book is that by teaching dogs ESL, you'll learn how to effectively use verbal instruction as a nonaversive alternative in any instances when physical correction or aversive means might be recommended.

Ultimately, for *your* dog's education, *you must decide for yourself* which training technique is best by asking the following questions of every technique: Does it work? How well does it work? How long does it take? How easy is it to grasp? And do you and your dog have a great time using it?

In the book's conclusion, I look to the future and speculate on what dog training might look like tomorrow with the help of advances in technology, increased research into dog training, plus *gamifying* the entire process of dog training.

The Accidental Dog Trainer

In many ways, I view myself as an accidental dog trainer. I never intended to become a dog trainer and dog behaviorist; it just happened. When I was five, my family decided that I should become a veterinarian because having a vet in the family would no doubt save money on the farm. But halfway through my stay at the Royal Veterinary College in London, I realized that opening a veterinary practice would stymie my peripatetic urges, and so instead I opted for research, which I had come to love. I was fascinated by obstetrics and simply loved behavior, and so I chose sexual behavior as a research topic. I tossed a coin to decide whether to research cows (my favorite beasties) or dogs. The Ugandan shilling decided dogs.

After enjoying veterinary college in London in the sixties, I decided to seek my PhD in the San Francisco Bay Area. I contacted Dr. Frank Beach at the University of California at Berkeley and received a reply that was more than I had ever dreamed of — a lengthy, engaging offer to join his thirty-year ongoing research study on the sexual behavior of dogs, automatic acceptance into a PhD program with Dr. Beach as my advisor, plus employment as a research specialist. My usual day comprised a couple of hours watching dogs "behave" in the California sunshine and then retreating to campus to analyze and share the results. I thought I had died and gone to heaven. I had always enjoyed watching animals, but researching dog behavior became a passion. Certainly, the seventies became one of the happiest decades of my life, even surpassing the sixties. Then quite by chance, it got even better.

The director of the UC Berkeley Extension contacted me about whether I would be willing to teach a ten-week course on dog behavior. I said yes, having absolutely no idea just how much fun it was all going to be. The registrants were largely dog owners with a few veterinarians, dog breeders, dog trainers, and shelter workers. They were spellbound. (I had never entertained such an enthralled audience since my debut onstage as a magician at the age of ten.) The dog owners ate it up, and the professionals bombarded me with questions... about *their* dogs. It has always amazed me that, even when I give academic talks at veterinary conferences, by far most of the questions are about a "client's" dog (yeah, right!) that chews, barks, soils, and otherwise wrecks the home. Dogs are just so special.

The ten-week extension course was primarily about behavior, but by unanimous request, I included a week on dog training — about which I thought I knew little to nothing. So I just chatted about how, when I was growing up on the farm, we lure-reward trained all animals off-leash, dogs included. The dog owners and vets imbibed the ideas, tried them out on their own dogs at home, and came back with glowing reports the following week. But the dog trainers were far from convinced and stared at me as if I were from a different planet reading from stone tablets. Some were skeptical to the point of contradiction, exclaiming, "Puppies can't be trained! You can't train dogs off-leash!" Total disbelief. Realizing that words would not convince them, I rounded up a few campus puppies and dogs (plus owners) and brought them to class the next week. The trainers loved it and were hooked. I was also hooked.

Back then, I remember thinking that I had the perfect "job" — educating, entertaining, and otherwise sharing my passion with others about developmental dog behavior and training. I simply loved lecturing, and so I started giving seminars all over the United States and around the world. I stopped keeping track of the number of one-day seminars and workshops over a decade ago when I surpassed fifteen hundred. I wrote a bunch of books, made lots of videos, started my own publishing company, started a video production company, and founded a puppy school, called SIRIUS Puppy Training, in the San Francisco Bay Area. I created and wrote the "Behavior" column in the American Kennel Club *Gazette* for many years, hosted a dog-training television program in England, founded the Association of Professional Dog Trainers, and bump-started many similar educational organizations around the world.

The whole time I just followed my nose. I planned very little. In the early 1980s, as I was starting SIRIUS, I remember driving across the Richmond Bridge to teach what was then the world's very first off-leash puppy socialization and training class and thinking, *Apart from a few dogs on the farm, I have never actually trained a dog myself. Hope it works.* It did. In 2023, SIRIUS Puppy Training celebrated its forty-first anniversary and now has over 120,000 puppy graduates and counting.

Nowadays, Kelly (my second wife) and my son, Jamie, pretty much run the business. During the Covid pandemic, we shut down all twenty-four branches overnight, but Kelly got SIRIUS Zoom puppy classes up and running within a few weeks, attracting registrants from around the world. For several years, Jamie filmed me on the seminar trail (in his quest to digitize his daddy's doggy brain), and hundreds of hours of these streaming videos now form the nucleus of Dunbar Academy (DunbarAcademy.com). All in all, Dunbar Academy represents a massive chunk of my life's work and probably the largest online resource for applied dog behavior and training.

My family's hankering for a vet of their own, a coin toss, dog behavior research at UC Berkeley, and a chance request to teach an extension course all pointed me in the right direction, but it was Mimi's "intervention" that sealed the deal. Mimi was the breadwinner as I renovated the house and traveled to give dog behavior seminars. One day in 1982, she cornered me during a weekend getaway in Bodega Bay and said, "I want you to take this cigar, go sit on that rock, and not come back until you can tell me how you're going to bring money into this relationship." It took just a couple of puffs: "I'm going to start a school for puppies, so that Omaha can go to school."

Since then, dog training has been my engulfing passion: refining reward-based training techniques for changing behavior and molding the temperament of dogs and other animals. To this day, my ultimate joy is teaching off-leash puppy classes. I love watching families thrill as their puppy's behavior and demeanor changes in leaps and bounds. As their confidence grows, all that endless puppy energy morphs into rapid recalls, quick sits, and rock-solid stays. The entire process is so easy and so predictable, yet for me, with each new class and each new puppy, it seems that *every time* the magic is happening *for the first time.*

I do so hope that after reading this book you, too, will experience the same magic with your dog.

Part 1

THE NATURE OF DOGS AND REWARD-BASED TRAINING

Dogs are dogs, and as such, they all grow up to act like dogs. Moreover, their doggy behaviors develop according to an utterly predictable developmental timetable. Often, owners consider many normal, natural dog behaviors to be problematic. Consequently, knowing that these behaviors will soon *predictably* emerge, it makes sense to *prevent* them from becoming problematic by providing appropriate and acceptable outlets for their expression *from the outset.* This involves teaching your dog *where* to eliminate, *what* to chew, *where* to dig, and *when* and for *how long* to bark or let off steam.

Species-specific and breed-specific genetic heredity exert huge influences on a dog's behavior and temperament. So does *social* heredity, or who the dog grows up with. For example, Rottweiler puppies grow up with a Rottie mother and littermates and Rottie people, and as soon as a puppy comes to your home, *you* become your dog's biggest influencer.

Your puppy's successful career as a companion, especially their confidence and friendliness toward people and other dogs, depends almost entirely on early enrichment, socialization, and training. The same applies to an adopted older dog. After just a few days with a new family, any dog starts to become a different dog.

As you progressively modify your dog's behavior and mold your dog's temperament, your dog becomes unique — one of a kind — *your dog*!

Part 1 overviews how nature and nurture work together to shape the personality of each individual dog, compares the different reward-based training techniques, and explains the secrets to using lure-reward training successfully.

Chapter 1

Dogs Are Dogs

Many years ago, I was swamped in a postseminar deluge of questions when a young fellow pushed through the throng and pleaded, "Please, I have a big problem." His need seemed pressing and so I asked, "What's the problem with your dog?"

He replied, "She barks!!!"

"Oh noooo!" I responded. "That's terrible. How unfortunate. To think of all the dogs out there, and you had the misfortune to pick one that barks."

Obviously, the young man's problem was not that his dog barked. Of course she barked. Just as cows moo, chickens cluck, and cats meow, dogs bark. In fact, barking is as natural and quintessentially dog as burying a bone or wagging a tail. The young man's problem was that he didn't like *excessive* barking or barking at *inappropriate times*.

Barking is surprisingly easy to control once you have taught your dog how you would like them to act instead (namely, to Shush on cue). Certainly, teaching your dog how to act is a whole lot easier than trying to suppress barking by joining into the fray and barking back, "For heaven's sake! Shuddup! Stop barking!"

Ahhh! a dog thinks. *Wonderful. A barking contest.*

As the young man exemplified, many owners find certain dog behaviors strange, annoying, or disgusting, such as drinking out of the toilet bowl and urinating on carpets. Doing these things doesn't make a dog "bad." Rather, dogs are simply good at being dogs. Almost all natural and expected dog behaviors can and should be allowed to have an appropriate and useful outlet. It would be unfair to ask a dog to live with us in our homes and then for us to try to suppress their normal, natural, and necessary doggy behaviors as if they were bad or wrong.

Lure-reward training is designed to teach dogs the appropriate times and settings for expressing their natural behaviors, so that their people never (or rarely) experience them as problems.

At the seminar, once the young man expressed his concern, it became obvious that many others in the assembled crowd had the same problem, and so I delivered a mini-seminar on the many fun ways to reduce barking and eventually get vocalization under temporal control by teaching Speak and Shush on cue.

The Power of Reward-Based Training: George

To give you a taste of what's possible with reward-based training techniques, allow me to share the story of George. On a previous visit, George's trainer, my partner Gina, had been treated to an altogether impressive performance — forty minutes of nonstop endurance barking at a single visitor. This made George a dog I was just dying to meet. Uncontrollable barking is my all-time favorite "problem," and so I asked Gina if I could tag along.

I always like to see a dog's behavior with my own eyes. I also like to quantify the behavior. First, so that I know exactly what I am dealing with, but also so that I can demonstrate improvement, especially the *speed* of improvement.

On my visit, George did bark for a good thirty minutes, but the barking was *not* continuous. It just seemed that way.

As soon as I arrived, it became obvious to me that George was barking for two primary reasons: first, the subject of this chapter. He was a dog, and it is not unusual for dogs to bark when they are bored, excited, upset, afraid, as a warning or threat, or because of a change in the environment, especially new people, other dogs, and other animals — or because a leaf fell from a tree three blocks away.

In this case, George was barking from excitement. *Visitors!* Also, his barking was being unintentionally reinforced by the owners. In fact, I think this is one of the best examples of unintentional training that I have ever seen — thirty minutes of nearly continuous barking. As I explained this to George's owners, they felt a mite downtrodden, and so I tried to put a motivational spin on it: "Well, I've never seen a dog trained so well, albeit unintentionally, to bark for so long with such enthusiasm, so we know that you all have tremendous training skills. Consequently, the prognosis for training George to bark less and Shush on cue is actually very good."

Always, my prime directive with any behavior problem is to reward the dog when they get it "right," that is, for doing the right thing in the right place at the right time. In this instance, my initial plan was to reward George for not barking. However, when I explain this approach to owners, many retort, "But I can't reward him for *not* barking. He never stops!" And so training never starts.

Absolutes, such as *always* or *never*, are rarely true. (I almost said never true, but then I thought of a slew of exceptions.) Long-duration behaviors are seldom continuous; instead, they are *episodic*. They start and *then* they stop. Some barking bursts may be lengthy, others short. Both the *onset* and *offset* of barking are marvelous educational opportunities, provided we give appropriate feedback.

George's training started by having everyone happily praise and reward him (with kibble, if he would take it) the very instant he *stopped* barking, if only to pause for breath: "Good boy, George. Woo Hoo! That's the way." With each pause, George received huge, jolly praise from all four family members, Gina, and me. Gradually, but progressively, over about five minutes, the periods of silence became longer and longer, from half a second to a full second, then two seconds, three seconds, four seconds, and five seconds. Yes, baby steps. But to go from half a second of inhalation to five seconds of silence is a tenfold improvement in training. That's a thousand percent! Wow! We kept going.

With longer pauses, it became possible to reward George the moment he stopped barking and for the duration he *remained barkless*. After another five minutes, we were getting occasional ten-second quiet moments. So we added in a nonreward mark, "UhhHrrrm!" — a sound like softly clearing your throat. Each time barking resumed, we all cleared our throat, turned away, ignored George, folded our arms, crossed our legs, and suspended all praise and food delivery.

After approximately fifteen minutes, we enjoyed our first thirty-second silent spell. At this point, George was ready to learn how to Speak and Shush on cue. Here's how: First, one owner instructed, "George, Speak." Next, I stood up, knocked on the tabletop, or did something silly to prompt George to bark. Then when George barked, all of us happily praised him and some of us joined in.

Each time we did this, the expression on George's face went from bemused to quizzical to excited delight, as if he were thinking, *At last! My people understand me.*

After half a dozen or so woofs on cue, we taught George to Shush on cue. First, one owner instructed, "George, Shush!" Then they waggled a food treat in front of George's nose. As soon as George started sniffing, he stopped barking, and immediately, all of us started happily praising George and offering pieces of kibble. (It is physically impossible to inhale air and bark simultaneously.)

Within barely a dozen repetitions, George began to reliably Speak and Shush on cue. George's barking, once so frustrating, annoying, and unstoppable, was now controllable, enjoyable, and offered many applications. For example, he could be prompted to bark three or four times and then shush whenever a nonfamily member drove up the family's extremely long driveway, parked, and walked up to the front door.

Our one-hour session with George ended with a record-breaking eleven-and-a-half-minute period of glorious silence.

In just one session, we made enormous strides with George's barking solely by *rewarding* George for four things:

1. Stopping barking of his own volition, if only for half a second.
2. Not barking (remaining quiet) of his own volition.
3. Barking on cue.
4. Stopping barking on cue.

The Dog's Point of View: Their Needs and Feelings

It should come as no surprise to anyone that dogs are social animals that thoroughly thrive on *communication* and *companionship*. Plus, they *live to sniff, revel in play*, and become immediately invigorated, refreshed, and refocused when allowed to be a dog and *let off steam*.

However, seldom do we establish clear communication channels with our dogs to teach them how we would like them to act, nor do we offer sufficient instructive feedback as guidance when their behavior goes off track. In fact, so much of *dog training has lost its voice*. Comprehension of verbal instructions is seldom tested, and meaningful, heartfelt praise seldom heard — often replaced by a click, or a treat, or no feedback at all. Many people take good behavior for granted and assume that dogs "misbehave" out of stubbornness or malice rather than because they lack confidence and comprehension.

Also, very few dogs receive the level of diverse and enriched companionship that they so desperately need, especially puppies during their formative months. Yes, puppies are loved by their breeder and new owners, but what puppies absolutely require is multiple opportunities to familiarize themselves with numerous *unfamiliar* people.

Years ago, my producers filmed a video for very young children in which I explained that scary dogs have feelings, too, and that most dogs are just as scared of us as we might be scared of them. I said this as the rationale for why people should stand still and never approach or reach out to an unfamiliar dog. My grandfather, a farmer, always impressed on me as a child that *to touch an animal is an earned privilege, not a right*. He and my father both taught me to wait for an animal to approach and make contact of their own accord. I wasn't to cavalierly go huggy-wuggy and kissy-face, or grab, restrain, and manhandle the dogs and cats or farm animals. They both considered an animal's voluntary approach and accepting a food reward as a revealing predictor and prognosis as to how the interaction would likely proceed.

As a youngster, I remember asking Gramp why I couldn't just hug cows, and his reasoning surprised me. First, he said what anyone would expect, "So you don't get bitten, kicked, gored, or pecked to pieces when you hug the wrong one." Then he continued, "Many animals are fearful of people because they have many good reasons to be fearful of people. So they are uncertain of your friendly, loving intentions. You must consider their feelings, proceed slowly, and learn to read each animal."

Both messages have stayed with me and influenced a lot of what I teach. They inspired me to *think dog* in order to understand what motivates them. I always consider, what do dogs really want? What pushes their buttons? What makes them the happiest? What are the best rewards?

Most of all, what dogs want is to feel welcome, safe, and loved in their home. They want to be included in day-to-day life, and special activities are a huge bonus. They want to be close to their people, both physically and emotionally. They are especially fond of seeing you first thing in the morning, and luxuriating with you for long, relaxed, chill times on the couch or in comfy dog beds in the evenings. They want to enjoy activities and games with you and savor long, lingering walks (and sniffs). Moreover, they dearly yearn for an education so that they may better understand their people and the ways of humans.

Above all, though, dogs want to *be dogs*, and so they need to learn *how* they can act like dogs in a manner that doesn't upset their people. They want to know *when* they can sniff the grass and investigate their surroundings, *when* to be active and silly, *when* it's OK for them to bark and howl, *what* to chew, *where* to eliminate or dig, and especially, *when* to jump up and hug and *when* to romp and play with other dogs. They desperately welcome instruction and guidance, so their play style doesn't annoy us, other dogs, and other people.

At the top of a dog's list of needs and favorite activities are companionship, freedom to sniff, lots of playtime, and a chance to let off steam. These are dogs' all-time favorite pastimes. They are so vital and engrossing that they are life essentials — any dog's reason for being. If these activities aren't incorporated into training, by being put on cue and/or used as rewards that reinforce training, then they will quickly become mega-level distractions that undermine training.

Companionship

Social animals, like dogs and humans, need to be social. They need regular companionship, otherwise loneliness and isolation will cause them to suffer mentally and emotionally. For dogs to adapt well to human society, for dogs to function without anxiety, stress, and fear of people in our crazy world, they need to be socialized as early in life as possible to as many *unfamiliar* people as possible. Open Paw's "Minimal Mental Health Requirements" recommends that puppies be handled, groomed, and trained by at least five unfamiliar people each day, that is, 150 *different* people during the first month in a new home. As a rule of thumb, I always recommend that puppies are handled and trained by at least a hundred different people in the breeding kennel prior to eight weeks, and then another hundred different people during their first month at home.

Safely socializing young puppies at home is as crucial as it is easy to do: Invite lots of people to visit and handle the puppy. However, as a precaution, have visitors leave their outdoor shoes outside to prevent possible infection from soil contaminated with viruses and bacteria.

A new puppy in the house is a wonderful opportunity to reenergize and normalize *your* social life. With a young puppy in residence, your house should be like Grand Central Station. A pup's special puppiness should be shared with family, friends, neighbors, and strangers. Don't keep your puppy a secret. Ask every visitor to bring along a couple of friends.

Naturally, wonderfully, your puppy develops the closest relationship with you and your family, and these special relationships are as important for dogs as they are for people. They also grow stronger as dogs grow older. But... the world is full of unfamiliar people, and you must *teach* your puppy to enjoy their company, too: veterinarians, groomers, people outdoors on public property, tall men with canes, women in hats wearing sunglasses, delivery personnel, children, children running and screaming, skateboarders, plus long-lost friends and distant relatives unexpectedly coming to visit. Otherwise, during adolescence, your dog will begin to fret and stress when encountering people they don't know.

Many families with a puppy are duped into believing that their new puppy is universally friendly, accepting, and confident, since this is how their puppy acts at home in a familiar setting, with familiar people, and other familiar companion animals. Then the family is shocked when their dog falls apart during a visit to a vet, in training class, or after encountering a stranger on a walk.

On my UK television series, *Dogs with Dunbar*, we once featured a German shepherd puppy named Shogun. The owners thought the pup was overconfident, overfriendly, over the top, and out of control, and even the TV show's researcher and producer were both ambivalent about accepting the pup for a televised, off-leash family puppy class. My two cents' worth: That's why we teach puppy classes — to locate problems and resolve them right away, and to modify and fine-tune puppies' personalities and temperaments.

Then, in class, Shogun spent the better part of the first two sessions hiding and shaking in the empty fireplace as the other eleven puppies took advantage of the ballroom venue to turbo-play. Shogun's owners were shocked. I asked them, "How many other shepherds do you have at home?" (Typically, shepherd puppy owners have other shepherds.) They said two, both male, and since Shogun played like a beast with his big brothers, they assumed Shogun was brimming with confidence. But big bold Shogun was now in a class with eleven *unfamiliar* puppies, and he was there *without* his two best friends and bodyguard big brothers, and so he fell apart. At the end of the second week, though, a shih tzu befriended Shogun and single-handedly provided the requisite puppy play therapy for a speedy rehabilitation. Soon, Shogun was playing with abandon with the other puppies.

The need for socialization with unfamiliar people is ongoing and lasts throughout a dog's life. When that happens, dogs progressively develop more and more confidence. On the other hand, they will progressively desocialize and lose confidence to negotiate the unfamiliar when they get stuck in a familiar social routine. A dog's temperament never stays the same. However, maintaining an adult dog's confidence and friendliness is easy and enjoyable. *A walk a day keeps fear at bay.* Simply stop every twenty-five yards and have your dog Beg, High five, and Playbow on cue for onlookers. Encourage people to meet your dog so your dog is always comfortable meeting strangers.

Perhaps the scariest people-related situation for puppies and dogs is *no people* — the infinite *nothingness* of those first few bedtimes *alone* in a new home or being left alone for lengthy periods during the day. Separation anxiety is agonizing for dogs, especially if, as puppies, they became accustomed to unlimited access to a vibrant family. Consider your puppy's point of view, and when you *are* home, teach your pup to enjoy short quiet moments on their own, chewing a food-stuffed chewtoy, in a dog crate or different room, so that your pup is better prepared to be confident and relaxed as an adult when left at home alone.

Freedom to Sniff

Dogs desperately need ample opportunities to sniff. All dogs want to use their nose to explore and savor an environment that is 99 percent hidden from us. A dog sniffs in vibrant, ever-changing, pulsing and flowing, 3D technicolor. In comparison, we sniff in moribund, myopic black and white.

My initial research at UC Berkeley comprised testing dogs' olfactory preferences by timing how long dogs sniffed different urine samples, such as from different sexes, neutered versus intact, young versus old, familiar versus unfamiliar, and so on. Simple preference testing offers a marvelous insight into a dog's brain. Then I averaged numbers, ran statistics, and calculated probabilities to determine whether dogs *significantly* preferred sniffing one type of urine over another. Duh! Yes, I know, I was proving the obvious, and as my grandfather used to say, "Could've told you that fifty years ago, boy." In my own defense, it's one thing to assume, suspect, or even know something; it's another kettle of fish proving it. Of course, dogs can tell the difference between two odors, but also, I wanted to find out what information dogs could discern from a few sniffs. For example, from sniffing a female's estrous urine, male dogs can determine the day of ovulation within twenty-four hours, which is certainly pretty useful for the preservation of the species.

However, after writing my PhD thesis, I read *Sirius* by Olaf Stapledon, a 1944 novel told from the perspective of a sheepdog named Sirius. Sirius's from-a-dog's-point-of-view writing changed my approach to research and to dogs in general. To quote Sirius: "But it's the smell (of her) that enthralls one, the maddening, stinging, sweet smell, that soaks right through your body, so that you can't think about anything else day or night."

After reading this quote, I compared it to a line from my doctoral thesis: "In a three-minute test, intact male dogs sniffed the estrous urine sample for an average of 13.3 seconds and the intact male urine sample for 1.7 seconds. ($P<0.01$ from a two-tailed t-test for differences between correlated means.)" The only doggy passion in my writing were the P values and the two tails. I was missing something, I realized. Something BIG. I hadn't a clue what sniffing meant to dogs and just how much dogs need to sniff. In fact, I was missing the whole shebang of being a dog! For dogs, odors are more than just numbers. Odors are everything. Dogs exist to sniff, and savoring sniffs *is* their *numero uno* raison d'être.

When Phoenix, my second Malamute, was a pup, I used to construct olfactoria for her. I would collect all sorts of odor samples, put them in metal boxes with a pinhole, and hide them in the drawing room: freeze-dried liver, lamb, dog urine (familiar and unfamiliar), Chanel N°5, cat poop, and pizza. I was fascinated by how quickly she located each odor source and which ones captivated her attention. Urine from her best bud, Oso, would keep her sniffing for ages. I guess she was trying to figure out what he was doing in a tin box. (At least he had a pinhole for air.)

Lots of Playtime

All types of dog play provide wonderful physical exercise that benefit a dog's physical health. Play increases bone density, exercises joints and muscles, keeps ligaments secure, and stretches tendons in a variety of directions. Play is far healthier for dogs (and humans) than straight-line jogging. Also, dogs are much more diligent than most people regarding stretching, with isotonic and isometric muscle contractions before, during, and after play. But aside from the many physical benefits, simply put, dog-dog play is essential for a dog's mental well-being and for mastering social savvy with other dogs.

Without regular off-leash play as puppies and adolescents, adult dogs often become dog-dog reactive and lack the confidence to engage. Also, they do not have sufficient opportunity to establish bite inhibition (via controlled play-biting), and so they are much more likely to cause damage even during minor disagreements. However, dogs that have learned bite inhibition in puppyhood almost never harm others during fights as adults.

Developing bite inhibition doesn't mean dogs are totally inhibited from biting. On the contrary, the *more* puppies play-bite with their needle-sharp teeth but weak jaws, the *safer* their massively strong jaws and bigger, blunter teeth will be in adulthood. Acquiring bite inhibition means learning to inhibit the *force* of their bites. Then, should they have an argy-bargy in adulthood, their larger teeth and stronger jaws will *cause no damage.*

Similarly, puppies must learn bite inhibition with people. First, they need to be trained to inhibit the *force* of their bites and then, and only then, to lessen the *incidence* of biting. If puppy biting is inhibited altogether, it is impossible for a dog to learn how to tone down the force and pain of bites.

Acquiring bite inhibition — from play-biting and play-fighting with other puppies, dogs, and people — is the number-one most important item on any puppy's educational agenda. As a rule of thumb, the more your puppy play-bites your hands, the safer their jaws will be in adulthood, provided you give appropriate feedback. If you are at all worried about your puppy's biting, or any other puppy problem for that matter, please read *Before & After Getting Your Puppy*, which offers nitty-gritty, how-to, puppy-raising advice, and prioritizes problem resolution in terms of urgency and importance.

The Bully Test

Unfortunately, lots of people inhibit free play, especially when dogs growl, play-fight, or have scraps, for fear that the dogs are trying to hurt each other. This is quite understandable because dogs snarl and growl during play in ways that resemble the real thing. If you are ever worried about whether your dog is playing or fighting "for real," try the Bully Test.

If your dog appears to be scared of or fighting with another puppy or dog, explain to the other owner that you are worried. Often, they are worried, too. To test, take hold of your dog's collar and see what the other dog does — run off and hide or run up to play? Usually, they clamor to get back to playing. They exhibit loads of atmosphere cues and gestures to advertise that they are friendly and to entice the other dog to reengage. To be certain, switch rolls; have the other person take hold of their dog and see what your dog does once free.

I frequently employ the Bully Test when teaching off-leash puppy classes to prove to owners that puppies are just playing. Once you recognize the signs of play — how both parties signal that they are friendly before engaging and repeatedly signal friendliness during every pause in play — you'll have the confidence to let them continue. If a puppy doesn't play enough during puppyhood, they won't learn how to act with other dogs as adults, they won't learn bite inhibition, and they'll likely be unwelcome on walks or in dog parks. Nonetheless, always interrupt play *frequently* to make sure that it does not amp up beyond your control.

A Chance to Let Off Steam

I remember a rainy day back in the early eighties. I was writing some articles for dog magazines, and Omaha was lying with his chin on the floor and his eyeballs moving back and forth, watching my every keystroke. I knew it was time for his afternoon run, but I was preoccupied, and it was raining quite hard. I heard Omaha let out a short *Grrwrroo*, then a few seconds later a much longer *Grrwrrrooorrrooroo*, and then he bunny-hopped twice, spun around and cleared the couch, blasted open a swinging door with his head, did two laps round the kitchen counter, charged through the still-swinging door, flew over the couch, and landed in a suction-pawed splat-sphinx as he rode the bunched-up, sliding carpet across the floor and into the bookcase under the window. His eyes were huge, and he looked possessed.

I called him and he came and rested his chin on my thigh, and I scratched him behind the ears. I felt justifiably guilty, and so I grabbed my coat, and we went out the front door. As I put on my coat, Omaha stood at the top of the porch stairs and looked up the street to the left. His nose twitched and he looked to the right. More nose twitching. Then he took a quick glance to the left, turned around, went back inside, and laid down on the crumpled carpet, letting out a long sigh and closing his eyes. He didn't want to run in the rain either, but he had become stir-crazy from being cooped up indoors all day. Half a minute on the porch and a few long sniffs were all that it took to relieve the pressure and calm the beast.

Most people interpret letting off steam as expending physical energy — a long brisk walk or letting a dog romp with other dogs in a dog park. They think tiring the dog to exhaustion will make them calmer and more relaxed. This often works in the short term, but in the long term, regular, extreme physical exercise becomes a form of *endurance training*. That is, it increases both the duration and intensity of activity required to tire a dog, such that the more exercise a dog gets, the more they yearn for more exercise. If calmness is your goal, then think about letting off steam in terms of providing *mental* exercise. Training games and especially sniffing require a lot of mental energy and both are more physically tiring than physical exercise.

After that rainy day with Omaha, I changed our routine. In addition to our afternoon eight-mile jog and our 1 a.m. hour-long neighborhood walk, we added short mental-health breaks every hour for just three or four minutes. We'd stand on the front porch, or check out the backyard, and then go back in. This prevented both of us from becoming antsy when confined indoors for long periods.

I started the same routine when driving on long road trips. I had never liked taking time-wasting pit stops, and I prided myself on never exceeding five minutes for car input and dog and people output. Increasing a stop to ten minutes made all the difference to the overall demeanor of driver, dog, passengers, and the trip as a whole.

Predict, Prevent, Provide, and Teach

It is not a mystery why so many people share identical dog "problems." Dogs develop only a limited number of behavior, training, and temperament issues, namely housesoiling, destructive chewing, excessive barking, digging, fear of people, biting, fear of dogs, fighting, hyperactivity, jumping up, pulling on-leash, and slow/no recalls. These twelve problems are so common that they comprise a good 87 percent of all dog problems, which makes it manageable to learn how to raise a good-natured, mannerly, and well-behaved dog by preventing or resolving all twelve.

More important than the obvious practicality of a problem-based syllabus for dog owners (and trainers), once you've learned *what's common*, behavior problems become utterly predictable. Moreover, the timing of their manifestation follows an equally predictable developmental timetable. Now, since dog problems are both common and predictable, it would be smart to prevent common and predictable problems in puppyhood, especially since prevention is so much easier, quicker, and considerably more effective than trying to break habitual problems in adulthood. As Kelly has always said: "Good habits are just as hard to break as bad habits, so teach good habits from the outset."

Here is the secret to solving problems:

1. Predict how a dog is likely to behave, both in general and at every stage of life.
2. Prevent mistakes around the house by not giving a new puppy or dog the opportunity to misbehave, especially during the first days and weeks at home and, especially, whenever left at home alone.
3. Provide appropriate outlets and ways for dogs to express their normal and natural dog behaviors in a manner that doesn't

upset you. Ask yourself: If what the dog is doing is "wrong," what is right?

4. Teach your dog appropriate and acceptable outlets for their natural behaviors, and praise and reward them handsomely.

My mantra has always been, if your dog's behavior is annoying you, how would you like your dog to act? For example, if you don't like your dog jumping on visitors, *how* would you like your dog to greet people? Maybe approach on their hind legs holding a silver platter with glasses of sherry? Or say to visitors, "Good afternoon. How do you do? Jolly pleased to meet you!"

Eventually, most people say, "Well, I guess he could sit."

"*That* is a really good idea!" I say. And I mean it. No joking around. The solution to the problem of jumping on visitors is to teach your dog that a quick sit and solid stay is the default greeting. So teach your dog to Sit-Stay on cue when greeting people at the earliest opportunity, and then motivate your dog to *want* to greet people in that manner. Then jumping up becomes history. This message echoes throughout this book, since it is the simplest solution for all misbehaviors.

Housetraining offers one of the best examples. Whereas many dogs appear to have limitless energy and can jump up, run around, and chew and bark for impressively long periods, they only have a finite supply of urine and feces. You could spend a lot of time and energy punishing a dog for *every instance* that they deposit urine in inappropriate places, but it's much easier, quicker, and more effective to predict *when* your dog needs "to go" and then *teach* your dog to eliminate on cue. How? By instructing your dog to pee in an appropriate toilet area and then handsomely rewarding them for doing so. Once your dog is delightfully empty, they have precious little urine for other areas of the yard or inside the house.

With puppies and adult dogs alike, after just three or four days of keeping to an hourly daytime toilet schedule, a good 95 percent of a day's supply of urine is deposited in the toilet area.

When housetraining puppies and dogs, I ignore their irrepressible cuteness and instead visualize them as a bulging bladder and turgid rectum — albeit camouflaged by a soft pet-able pelage with larger-than-life eyes — that both require hourly emptying. When I praise and reward puppies for promptly peeing and pooping on cue in the appropriate spot, many people think that I've lost my mind. I sing and dance a jig (which is helpful if they still need to poop), while offering several sequentially fed treats. I like to make a statement. Yes, I am happy that my dog *did it on cue and in the right spot.* But I am also overjoyed that they are now, temporarily *empty.*

And let's just consider what the dog is thinking, *Wow! If only they'd let me know that I could cash in my urine for treats, I wouldn't have wasted so much on the carpet.*

Instead of inhibiting dogs from acting like dogs, teach dogs *how* to act like dogs *appropriately* for a domestic setting. Teach dogs *when* and *where* to eliminate (on cue in a doggy toilet area), *what* to chew (chew-toys), *when* and *where* and for *how long* to bark (by teaching Speak and Shush on cue), *where* to dig (digging pit), and especially, *how, when, where,* and *for how long* to be hyperactively silly and let off steam.

Training need not be a chore. Indeed, it should not be a chore. Rather, approach training as the way to build a caring, kind, loving relationship with your dog. At first, your most pressing goal is to teach basic household manners, so your puppy does not annoy you. Then practice cuing both the onset and offset of your dog's nuclear outbursts. I always feel so sad when I see owners become frustrated or angry because of their dog's occasional, or frequent, happy outbursts of barking and hyperactivity.

With lure-reward training, living with a dog becomes the joy that it should be.

Chapter 2

But There's No Dog Like Yours

Yup, dogs are dogs. Dogs look like dogs, act like dogs, and think, feel, and learn like dogs. Even so, every single dog is unique — a one-off. There's never been a dog like yours and never will be. Your dog is special.

What makes each dog unique? A combination of genetics and experience, nature and nurture. Genetic heredity comprises natural selection, selective breeding, and sexual reproduction. Experience comprises social heredity, early environmental enrichment, socialization, and training. With domestic dogs, people have total control over most of these. Through selective breeding, people influence genetics, and a dog's upbringing is entirely in people's hands from birth through old age.

That said, by the time most people get a pup or dog, the Gene Team has made its play, and whatever experiential enrichment, socialization, and training occurred beforehand, whether in a breeding kennel, another home, or a shelter, and whether those experiences were up to the mark, subpar, or missing — is now water under the bridge. The only thing that a new owner can do is *socialize and train their dog now* — to give their dog confidence to thrive in their new home and the world of people.

Without a doubt, regardless of breed, breeding, and past experiences, ongoing *socialization and training are the two most* important *variables* that will influence and determine the development of your dog's behavior, temperament, manners, and quality of life, for the rest of your dog's life. A dog's personality and temperament are not set in stone. They change over time, naturally with development, but more so by experience, ongoing socialization, and learning from you.

Certainly, handling, socialization, and training are best started in puppyhood. I've always maintained that *the puppy is parent of the dog*, that is, the nature of puppy socialization and training molds and solidifies adult behavior and temperament. While you have no control over the past, *you* hold the keys to your dog's future quality of life.

A Solid Temperament Is the Foundation for All Training

If puppies are not exposed to a wealth of diverse stimuli and numerous unfamiliar people when very young, as they grow older, they will *naturally* start to become increasingly wary of unfamiliar people, handling, strange noises, and so on. Surprisingly quickly, wary dogs become standoffish, then shy, then fearful, then overwhelmed, and then maybe reactive. At five to eight months of age, all it may take is an unfamiliar person, a child reaching for the dog, a veterinarian examining the dog's ears or paws, a tall man with a beard, a sudden movement, someone with a strong perfume or an unusual gait, someone falling down, the hiss of a truck's air brakes, or a thunderstorm, and a poorly prepared dog might go into a full-blown panic. Seemingly out of the blue, a dog has an exaggerated fear reaction to what should be a familiar and nonthreatening stimulus. When panicked, dogs often become unresponsive to instructions.

All obedience training, all that flashy heeling, all those lightning recalls, quick sits, and rock-solid stays, can crash and burn in an instant if an adolescent dog lacks confidence. A single unfamiliar dog, or a child on a skateboard, can destroy a dog's attention in an instant, sending it down the drain, along with all their learned responses.

Just recently, we had an unexpected dog arrive at our house at 10 p.m., and so I went outside to get an extra crate. When I returned, I came out of the dark, bent double with a forty-two-inch metal crate on my back. Sam, our extremely well-socialized, one-year-old Bernese, went entirely ballistic. I stood still and talked to her, holding the crate steady on my back with my left hand and offering kibble from my right. The other dogs readily took the kibble while Sam continued to bark. It took about fifteen seconds before she tentatively approached and sniffed me (to confirm my identity), and then she, too, accepted kibble. Just to be certain, I did a couple of laps around the room, still bearing the load on my back, and Sam happily heeled by my right side (the food-delivery side), munching her late-evening snacks. That was unexpected. Sam has seen me carry crates many times before. The takeaway from this: The problem was identified *and resolved* in less than a couple of minutes.

Desensitizing your dog to every possible setting or scenario and socializing your dog to all types of people as well as to other dogs and animals is vital and overwhelmingly important. No matter how much you try, though, there will always be unexpected cracks in your dog's temperament. Resolve them right away.

Prior to eight weeks of age is the best time to socialize puppies to people and to expose them to the myriad of normal yet unfamiliar stimuli in the world of humans. The most wonderful advantage of very early socialization is that it can be done without causing undue stress. The second-best time is before twelve weeks. The third-best time is before eighteen weeks of age, which is the beginning of adolescence, when dogs start to become wary toward unfamiliar stimuli. And the fourth-best time is *right now*!

Nature and Nurture: Genetic and Social Heredity

For several hundred years, scientists have debated: What is more important for behavioral development — nature or nurture, genes or experience? For quite some time, there was a tremendous bias toward genetic heredity as the main determinant of behavior and temperament, but the past century has produced a wealth of evidence illustrating the many ways that experiences change behavioral development, especially early experiences.

So, with our dogs, which is more important? Do we *breed* good dogs or do we *raise* good dogs? Of course, this isn't an either/or question. It probably is not even a question at all. Rather than wasting time and effort quibbling about the relative importance of genetic versus experiential factors, we should focus our efforts on ensuring that only the very best genes are passed along to the next generation and that puppies receive the best of all upbringings.

Most important, we must realize that the instant a spermatozoon has wiggled its wily way into an egg, creating a zygote, the only way to influence and guide the development of a dog's behavior and temperament is through *environmental stimulation*, starting in the uterus (good exercise, good health, good food, and good fluids for the mother), neonatal handling, ongoing handling, and *safe* socialization and training with unfamiliar people in unfamiliar situations throughout puppyhood and adolescence, plus socialization with unfamiliar dogs as soon as it is safe to do so.

Genes Impact the Senses

Genes affect a dog's senses, and even minor inherited differences in hearing, vision, and smell can have dramatic effects on how a dog perceives their environment and reacts to environmental change. These differences affect behavior, working ability, and ease of training.

For example, shepherd dogs were selectively bred to be extremely sensitive to sudden changes in the environment, especially auditory and visual stimuli, and to have the curiosity to check them out. Similarly, obedience breeds were selectively bred to be sensitive to human movement and voice, which means that, by and large, they are easy to train.

However, there are pros and cons to sensitivity. If "sensitive" breeds grow up in a limited environment during puppyhood, as adults, they often become oversensitive and overreactive to the presence and actions of unfamiliar people and to sudden changes in the environment. Sensitive breeds need to be mega-socialized during puppyhood, and desensitized to all sorts of noises and sudden movements by children, men, and strangers, so that as adults, they don't overreact to environmental change and unfamiliar people. An oversensitive puppy will *not* "grow out of it" *without your help*. Instead, fear characteristically intensifies with age. And so, immediately desensitize any incipient signs of anxiety or stress.

All puppies need environmental enrichment, handling, and socialization with lots of unfamiliar people as early as possible. Yes, all novel stimuli and situations can be a teeny bit stressful for a young pup, but this is nowhere nearly as stressful as what can become adrenal-emptying, overblown reactions to fairly innocuous stimuli in adulthood for dogs that were *not* exposed to a wide variety of stimuli and handled by many people as puppies. We must expose pups to every possible stimulus and scenario that they might possibly encounter as an adult, so that *the unfamiliar of puppyhood becomes the familiar of adulthood.*

Upbringing Trumps Breed

People generally think that every breed has a distinct temperament, and that breed determines how a dog will behave. Yes, we've bred dogs that behave extremely differently from other breeds, just as breeds look different from other breeds. Consider the differences between Labradors, German shepherds, bulldogs, beagles, bassets, Bernese, greyhounds, Rottweilers, Malinois, Malamutes, dachshunds, Great Danes, shih tzus, and Chihuahuas. Based largely on their breed-specific genes, we would never expect a Malinois to behave like a Bernese, or a basset to mimic a greyhound.

However, a puppy's social upbringing also has a significant influence, that is, its breed-specific culture. Young puppies grow up with Mum and their littermates, and their early social environment — *social* heredity — has a massive effect on their behavioral development. For example, golden puppies are raised with other golden puppies in a golden environment with a golden owner, so it's not surprising they grow up to act like goldens. The same as when bull terrier puppies are raised with other bull terriers in a bully environment with a bully owner.

But what if we conducted a cross-fostering experiment? Let's say we placed a single golden puppy in a "let's-get-it-on" litter of bull terriers, and placed a single bull terrier pup in the sun-is-shining, fetch-this-fetch-that, just-wag environment of a golden litter? Which would win out: genetics or upbringing?

I know of several anecdotal "cross-fostering" scenarios in real life. For example, a friend of mine living with an elderly female Yorkie-Chihuahua mix welcomed a young male Great Dane puppy into their household. Just thinking about it cracks me up. The effects were huge. The male Great Dane grew up adhering to Yorkie-Chihuahua rules and to Yorkie-Chihuahua customs and culture, and the Great Dane puppy became a Yorkie-Chihuahua at heart — a yappy lap Dane.

Additionally, people have an *enormous* influence on a puppy's breed-specific social environment. Border collies and bichons grow up with border collie and bichon owners, and typically, owners who favor different breeds have wildly different views regarding rules and their pup's educational agenda.

Moreover, some puppies grow up with a single person in a small apartment with few visitors; others grow up in large families with lots of children and frequent outings and gatherings for birthday parties, picnics, and so on.

It is truly eye-opening to see two sibling puppies when they meet again in puppy classes. In Puppy 1, they were like two peas in a pod, but just five weeks later in Puppy 2, they are like chalk and cheese. It's always a *Tale of Two Dogs*. The differences between the two puppies are a hundred times greater than the differences between any two breeds, especially in terms of their behavior toward their owners, other people, and other dogs. One might be asocial and fearful while the other is friendly and brimming with confidence.

A breed stereotype predicts how a dog would likely grow up *all things being equal* and *without human intervention*. But things are never equal, and of course, people intervene. Your puppy's early environment, social experiences, and education are by far the most important factors that influence and determine their adult manners, behavior, temperament, and personality. And you have total control over these experiences.

For the first few weeks at home, a puppy is almost a blank canvas with just the outlines of a breed stereotype. Then *you* paint the picture as you raise this puppy. From day one, your puppy's brain, behavior, and temperament will develop differently than if they lived with any other person. Your puppy is *your* legacy.

Remember, training isn't just about teaching a dog to Sit on cue. Puppy training focuses on molding temperament so that, as an adult, your dog can handle with aplomb whatever life throws at them.

Environment Changes Brain Anatomy

The impact of experience is so powerful that it changes the brain's anatomy. This was shown in a series of experiments by Drs. Marion Diamond and Mark Rosenzweig in the 1960s. I was both amazed and fascinated when I first read about these experiments when I was at the Royal Veterinary College, and once I joined Dr. Frank Beach's dog behavior research program at UC Berkeley, I made a point of meeting Mark Rosenzweig (three offices down from Dr. Beach) and Marion Diamond. Since I had studied neuroanatomy in school and at college, Dr. Diamond invited me to be an informal (unpaid) research assistant.

Up until the 1950s, it was thought that the structure of the adult brain was immutable, fixed in stone, and resistant to change. However, Drs. Rosenzweig and Diamond and their colleagues proved otherwise. They compared rats living in standard cages to rats living in cages with lots of rat toys, tunnels, ladders, and running wheels. They demonstrated that increased sensory stimulation from environmental enrichment caused changes in brain anatomy *within just three weeks*. Brain changes included a thicker cerebral cortex, larger neurons, increased dendritic branching, more dendritic spines, synaptogenesis (more connections between cells), and increased cholinesterase (larger/faster synapses). In terms of brain function, the animals showed better learning, retention, and problem solving. The scientific community was stunned.

And which enrichment "stimulus" had the greatest effect on brain anatomy? Another rat! That is, the social stimulation from hanging out and cohabitating with a ratty roommate!

Subsequent studies showed that the effects of environmental enrichment and especially social stimulation were even more dramatic the younger the animal. For all newborn animals, humans included, handling and socialization are crucial and of utmost importance to stimulate *normal* neuronal development.

Early environmental enrichment alters the expression of many genes, which have huge effects on determining neuronal structure and function, behavior, temperament, learning, and memory. The adult brain certainly has a lot more neural plasticity than previously thought, but nowhere near the brain changes that occur in neonates and youngsters.

However, the process works both ways. The researchers also showed that the beneficial changes were reversed after just three weeks of environmental deprivation. In other words, *use it or lose it!*

Environmental enrichment and social stimulation are essential for neonates and young puppies, since both *stimulate* a postnatal explosion of synaptogenesis — brain cell growth and branching, which establishes new connections (synapses) — and both help *prevent* later apoptosis (death and radical pruning) of millions and millions of unused or underused brain cells.

Dogs Have Multiple Personalities

Many people often underestimate the complexity of their dog's personality. Because of genetics, sexual reproduction, upbringing, environment, socialization, and training, every dog truly is unique. They have personalities as nuanced as our own. People often describe their dog with a single adjective, like affectionate, loving, faithful, stubborn, willful, naughty, hyperactive, and so on. However, just like us, dogs have numerous personalities based on the occasion.

For instance, years ago, the director of my UK television series got a male Labrador pup that he "coincidentally" named Dunbar. The pup was always around the set, and this caused considerable confusion when the director would instruct his dog, "Dunbar, Come," "Dunbar, Down," and "Dunbar, Off." I frequently used Dunbar as my demo dog on the show, and to prevent confusion, I decided to call him Dumpy. As a young adolescent, Dumpy produced some remarkable footage. We took him to a variety of safe off-leash areas in the New Forest and let him roam, while I and the crew hid in the bushes and filmed him from a distance, as I gave a hushed, Attenborough-esque commentary. Every time Dumpy encountered a different dog, he assumed an entirely different personality, ranging through the whole gamut of emotions: playful, pushy, unsure, amorous, over-the-top, low-key, timid, and so on. Dumpy's enormous thespian range truly surprised me. The multifaceted Dumpy could play any role.

All dogs have personal preferences, and it's certainly worth paying attention to your dog's preferences, since dogs are often less tolerant with people or dogs they don't prefer. Initially, puppies prefer children to adults and females to males. After a short while, the pup's favorite person in the household is usually the "caretaker," not just the hand that feeds but also the person who takes responsibility for housetraining and joins the dog for walks, training, and play. May I present the human adult female in the family? There are exceptions, of course. Puppies sometimes bond quicker and more intensively with girls, boys, or adult males. Indeed, it is not unusual for an adopted dog to be an obvious "man's" dog.

Just like Dumpy, you'll notice that your dog acts differently with different dogs and people. All dog trainers hear the line, "He's a totally different dog when he's with you."

As an aside, sometimes people worry that, when there's someone in the house who fails to follow established training "rules," their lack of cooperation might ruin the dog's training. However, that seldom happens. Dogs are much too smart to let a slacker ruin everyone else's efforts at communication and education. Dogs consider who said what, and they act differently according to the specific individual. So if *you* train your dog, then *you* will have a trained dog. A dog's misbehavior with someone else rarely affects the dog's compliance and engagement with you. Plus, the more your dog engages with you, the deeper the relationship.

All of this holds true with how dogs relate to other dogs. Dogs have dogs they like, dogs they dislike, special friends, and special enemies.

This is something to consider whenever you decide to add a new dog in the house. I often ask people if they have let their dog have a voice in the decision. Some people with a single dog believe their dog must get lonely and want a doggy companion, but do they? Maybe they do. Maybe they don't. Or maybe they would like to have a say in the choice of companion. I like to invoke a human analogy. What if someone said to you, "You know, I think you must be lonely and bored living alone, so I've asked a homeless man to come and live with you in his forever home."

When adopting a new dog, always make sure that your resident dog agrees. I remember when we adopted Claude, a 110-pound fox-red Rottweiler-hound cross, the SPCA would not allow us to bring Olly, our resident medium-sized, color-matched couch-and-bed mixed breed to check him out. Nor would they allow us to take Claude outside the facility to meet Olly. The director was a good friend of mine, and I explained it was not cool fundraising optics to refuse adoption, especially when Claude was scheduled to be euthanized the very next day for "biting" a shelter employee, and the facility was billed as a no-kill shelter. The director offered his office for the introduction. Olly was not overly enamored with Claude, but he didn't seem to object that much. So Claude became Olly's new roomie.

Molding Temperament and Creating Confidence

Behavior may easily be changed (that is the basis for this book), and changing behavior indirectly changes temperament. Simple things like teaching a dog to sit and watch you (rather than eyeballing another dog), and to act friendly on cue (such so to playbow), helps them to *feel* calmer and friendlier. Also of course, both behaviors change the perception of *other* dogs and owners, causing them to feel less threatened by your dog, and so they threaten your dog less.

Similarly, we can mold temperament (also the basis for this book), and changing temperament directly changes behavior toward unfamiliar people and other dogs. For example, asking other people to practice repetitive, very short Come-Sits is the quickest way to teach a dog to trust and eventually enjoy being close to them, which obviously reduces shying away, barking, and other signs of fear.

One day as a child on our farm, I remember seeing all eighty young heifers in the farmyard. They were filing one by one down a walkway between the pig sties and a cattle pen. Each heifer walked into a cattle crush, where she was given a bunch of silage, while Mr. Brinklow, a farmworker, fondled her ear and scratched the side of her neck, whereupon the heifer was released to walk back to the pasture, and the next heifer walked into the crush. I asked Gramp what was going on. He replied, "Vet's comin' t'morrow. At least the girls will know what to do and won't give vet any trouble."

The vet would be touching the cows in the exact same way, checking the heifer's number tattooed in her ear and measuring the lump on her neck, or lack thereof, from the tuberculosis test. Aside from the fact that Gramp wanted both the vet and his heifers to feel at ease, he knew that unmanageable cows would incur a larger vet bill.

After qualifying as a veterinarian, I realized that hard-to-catch, fearful, aggressive, intractable, and even wriggly animals are the biggest time-wasters and stressors in veterinary practice. This is true for cats and dogs, pocket pets, parrots, horses, and exotics. When animals are easy to handle, examination and treatment are considerably easier, quicker, more effective, and more enjoyable for veterinarians and their patients. When an animal is sick or injured, and in discomfort and pain, nothing could be worse than being restrained in an unfamiliar clinic environment and examined by an unfamiliar person. For many animals, the mental anguish hurts more than the physical pain. Confident animals, on the other hand, do not require physical restraints or time-consuming, stress-free handling techniques. Instead, not feeling stressed, they happily let vets do their job.

Socialization and Handling for Puppies

Changing behavior and changing temperament go hand in hand: Changing behavior — for example, by teaching your dog *how* to act in stressful social situations — imbues confidence. Similarly, building confidence via classical conditioning, desensitization, handling, and socialization during puppyhood creates dramatic and long-lasting changes in behavior.

When we use the words *enrichment* and *socialization*, most people tend to think of letting a few people meet the puppy or giving the puppy a couple of toys. But sensory stimulation, exploration, petting, play, and training have immediate and indelible effects on brain development, so during the first four months or so of a puppy's life, it's crucial to provide intensive puppy handling, mega-socialization, and beyond-your-imagination environmental *enrichment*, comprising a circus-pantomime cacophony of novel paraphernalia, and especially lots of children. As puppies get older, this includes regular opportunities to explore and play with other dogs, and most importantly — a walk a day for the rest of their lives.

Birth to Eight Weeks

Many dogs are born and live in kennels until eight weeks. Still, even during this time, it is vital that numerous people, especially men and children, handle and train the puppies. Neonates and young puppies need to be cuddled, stroked, and petted every wake/sleep cycle. As soon as Mum has licked a pup's anogenital area to stimulate reflexive urination and defecation, pick up a single neonate and stroke their closed eyes and ears; handle their teeny paws, muzzle, belly, and inguinal area (once emptied, unless you would like a garnish of dripped urine); and cradle and cuddle them in your hands, prone and supine. Let the neonate breathe your breath, and then do likewise with another pup.

Although very young puppies can neither see nor hear, they can feel and smell, and so they can identify different people. Early puppyhood is the very best time to accustom puppies to being hugged and handled by unfamiliar people and to desensitize subliminal bite triggers. There is little difference between being hugged or restrained, or between being petted or examined, other than the dog's perception of the handler. Neonates and young puppies are all about handling, handling, and more handling.

The first week after the neonates' eyes and ears have opened is the best time to expose puppies to sights and sounds. Play tapes of all the sounds and noises that often frighten adult dogs. Start with a very low volume and increase it gradually day by day. Most important, hold up each puppy to stare into their eyes, to accustom the pup to human eye contact. Hold each puppy in both hands so that the rear quarters dangle, and then lay the pup supine in your lap for a gentle and rhythmic belly-and-chest rub. Mesmerize the pup so they relax like a rag doll and *love* being massaged. Rub a finger inside their ears, open their muzzle to feel their teeth, and gently examine each digit of their paws, so that when the pups grow up, the owners will be able to clean the pup's ears and teeth and clip their nails.

Everything is new to a puppy, so expose them to everything. Once pups are four weeks old, build a puppy play/activity room. Cover the floor with every conceivable substrate, bubble wrap, crinkly paper, cardboard boxes, step-towers of training platforms, wobbly plank seesaws, and children's plastic swimming pools brimming with plastic-colored balls, loads of different dog toys, and especially lots of chewtoys to begin chewtoy-training because the young puppies will try to chew *everything*. Puppies consider every insentient object to be their chewtoy. This way, nothing will be unfamiliar and give them the heebie-jeebies as adults.

Eight to Twelve Weeks

For most puppies, this is their first month in their new home, and the most influential and important aspect of the environment is people, especially you and your family, friends, and neighbors, that is, the people that your puppy is likely to encounter on a regular basis when an adult. Ensure all family members regularly repeat the handling exercises, lure-reward train basic manners, and practice toy exchanges (for kibble) to prevent protection of valued objects.

Next, invite friends and neighbors. Having neighbors meet and like your puppy can be very useful, for example, for when you need someone to pop in to let the puppy out for a toilet break or feed them dinner if you're going to be late getting home. And with dog-owning neighbors, maybe plan to exchange occasional dog-sitting duties.

During the first month at home, make sure your puppy meets and is handled and trained by at least a hundred *unfamiliar* people. This is easily doable if you host a couple of Puppy Parties each week. I have received many cards and emails from people who have read my books and wanted to let me know that they surpassed the suggested minimum of a hundred unfamiliar people within just one week and that their dog now has a golden temperament. One Malamute owner exceeded a hundred people in just four days. When Kelly and I picked up Dune, our American bulldog puppy, at one of my seminars in Pennsylvania, he met over a hundred people in one afternoon and then more on the flight home, including the entire flight crew.

Invite your friends and neighbors to bring along some of their friends to each Puppy Party. Make certain that every guest leaves their outdoor shoes outside. At this age, acquired maternal immunity is rapidly declining and immunity from vaccinations is not yet fully established.

Ask your guests to dispense the pup's entire dinner of kibble, one piece at a time, to classical condition and imprint on your pup's brain that people are awesome! Instruct all visitors to take hold of the pup's collar and look in their eyes. Then, while handfeeding kibble piece by piece, they should handle ears, muzzle, paws, and nether regions and give a big hug and chest scratch. Then teach everyone how to teach Come-Sit, and have them repeat one-step Come-Sits over and over to accelerate bonding and acceptance, so this becomes your adult dog's default greeting.

Be creative. During subsequent parties, introduce novel, unexpected, and unusual situations. Have people dress up like it's Halloween and wear dark glasses, hats, all sorts of masks, and different aftershaves and perfumes. Ask people to bring unusual objects, such as umbrellas, skateboards, and noisemakers. Have everyone act weirdly. Adult dogs that have not seen it before find it extremely spooky when people walk with unusual gaits, fall down, and cry, especially children. Without overwhelming your puppy, prepare them for anything and everything.

Make sure that a family member monitors your puppy's reactions and demeanor *at all times* during the party. If the pup is happy and not worried, wonderful. Carry on. However, if the puppy becomes unsettled, just instruct, "Everybody, stand still." Then have each person call the pup to Come-Sit a few times.

A pup's first month at home is the ideal time to go wild with lure-reward training to expand their vocabulary. Navigating rapidly approaching adolescence is so much easier for puppies if they already understand what you are saying and what you want.

Finally, start to introduce your puppy to the world at large. You can safely socialize a young puppy in public by carrying, carting, or driving them to well-populated places, such as along Main Street or to mall parking lots and large stores.

Twelve to Eighteen Weeks

As your puppy meets dozens and then hundreds of new people — safely, both at home and out in the world — all were once strangers, but now are visitors, friends, and pleasant memories. Sit on a park bench and give your puppy ample opportunity to settle and watch. As they spend hours upon hours outside soaking in the sights, sounds, and smells, and especially watching other dogs, they become familiarized and unbothered by the ever-changing hustle and bustle of daily life in our overstimulated human world.

During twelve to eighteen weeks, continue having Puppy Parties, but also focus on establishing a core social group of dogs — your family's, friends', and neighbors' dogs as well as pups from off-leash puppy classes. Long-lasting friendships are often forged in class, and these can form a wonderful nucleus for your puppy's core group. Going on walks and park trips is much more fun in a group. Moreover, a group of close doggy and people friends offers the best therapy should your pup have occasional scary social experiences.

Off-leash puppy classes provide a wonderful, safe, and controlled setting for puppies to socialize with people, especially children; to acquire dog-dog bite inhibition; to socialize and play with other puppies; and to learn dog savvy. And this happens all under the watchful eye of a trainer on the lookout for incipient temperament problems, such as backing away from people, hiding, or "bullying." In class, these problems may be resolved as soon as they are noticed.

Additionally, you will learn off-leash control in a safe, albeit extremely distracting, setting, which will make your puppy much easier to control when off-leash *at home* and on walks and in parks, where your dog may *continue* to meet unfamiliar people and other dogs.

Myths and Fears about Early Socialization

A surprising number of people think early socialization is unwise, dangerous, or unnecessary. As a result, puppy socialization, handling, and training are too often seriously neglected. Consequently, utterly predictable and easily preventable temperament problems start to emerge in older puppies and young adolescents. Here are my responses to some of the main reasons people give for avoiding early socialization.

Myth: Unvaccinated Puppies Risk Infection

Some veterinarians advise not to go to puppy classes or walk puppies until they have completed their full series of immunizations. Of course, socialization must be safe for puppies, and the risk of infection is a real concern. So wait until your pup has received their full quota of immunizations before walking them in public places and neighborhoods that might be frequented by unvaccinated dogs, and hence might be contaminated with infected urine and feces. Especially, when visiting the vet, carry your puppy directly from your car to the examination room table. Don't let puppies walk through the clinic car park or lie down on the waiting room floor.

However, your puppy also requires urgent socialization to "vaccinate" them from later temperament problems. To socialize your puppy safely in public, don't let them walk on or sniff the ground. Instead, carry, cart, or drive your pup to places with lots of people and encourage others to greet your pup through the window in car parks.

When people come to your house, as noted above, ask them to leave their outdoor shoes outside to prevent fomitic infection from infected soil. During Covid times, we also had to make sure that puppy socialization was safe for people — so we instructed visitors to wear masks and meet pups outside at the end of a six-foot leash.

Myth: The "Fear Imprint Period" Might Harm Puppies

One of the best-known and most-often-quoted aspects of dog behavior is the "fear imprint period," which supposedly arises at the seventh week (some even claim on the forty-ninth day), but this period simply doesn't exist. There is no scientific evidence for a "fear imprint period" in puppies. This was merely an opinion of researchers of a single badly done study that involved shocking young puppies as they attempted to approach the experimenter's assistant. Some puppies became fearful. No wonder. Additionally, these puppies were raised in kennels with next to no interaction with people. Pure silliness. The suggestion of a "fear imprint period" has been used to abruptly inhibit, pause, or sometimes curtail essential handling and socialization exercises at a highly impressionable time in the puppy's life, which then causes the predictable development of fearfulness in many adolescent dogs.

Myth: Socialization Is Too Stressful for Puppies

It is next to impossible to overstress neonates and young puppies by handling, hugging, stroking, or petting even by numerous strangers. Puppies less than eight weeks of age are stimulus sponges. When overstimulated or overwhelmed, young puppies simply fall asleep (to solidify the experiences in long-term memory). When they wake, they yawn, stretch, eliminate, suckle, and then play, play, play until they're physically or mentally exhausted and collapse on the spot, only to wake up raring to go once more.

Any novel stimulus in puppyhood causes very mild and short-term stress — as evidenced by tiny blips in corticosteroid levels. However, puppies that were handled regularly as neonates and very young puppies *do not experience* the massive, overwhelming, and disproportional corticosteroid surges experienced by dogs that were seldom handled as puppies. Minor stressors in puppyhood enable better coping with mega-stressors during adulthood. On the contrary, it is a lack of early handling and socialization that condemns many puppies to a miserable life of crippling and insuperable stress and anxiety as adults.

Myth: Confident Puppies Don't Need Socialization

Socialization is so deceptive. People are sometimes duped by their puppy's confident and friendly demeanor — at two, three, or four months old, puppies are usually overconfident, overfriendly, and easy to handle. They're puppies! *All* young pups are universally outgoing toward people. Fear of unfamiliar people and dogs, and of novel stimuli and situations, doesn't emerge until much later during early adolescence. This means that breeders, veterinarians, and owners are usually unaware that anything is awry. Consequently, people are shocked when their adolescent dog gradually but progressively becomes wary, fearful, reactive, and maybe aggressive.

The early warning signs of future fear and aggression are often subtle and easily missed. You must always be on the lookout if your pup is ever slow to approach certain family members, ducks their head and backs away, or freezes or struggles when held. Resolve these developing problems immediately.

Some owners consider struggling when restrained an annoyance or the sign of a "dominant" dog that requires teaching "who's the boss." Other owners notice warning signs but do little more than euphemistically explain them away: "He's a bit hand-shy." "She takes a while to warm to strangers." "She was never really fond of my husband's friends." These are not little issues. For the puppy, being shy, scared, anxious, or stressed is an ongoing worry that won't go away until an owner does something about it. Most important, if the pup is not play-biting, they cannot develop bite inhibition, and so any fights or bites as an adult will be more injurious. Resolve the problems immediately.

Adolescent-Onset Fears and Anxieties

Some behavior and temperament problems catch owners by surprise, such as fear of unfamiliar people, fear of unfamiliar dogs, fear of unfamiliar stimuli and situations, and separation anxiety. However, adolescent-onset temperament problems are as predictable as chewing, barking, burying a bone, or wagging a tail, since they all stem from the *normal* course of social development. They are to be expected in all dogs as they transition from puppyhood to adolescence, *unless we intervene*. Early puppy handling and training by numerous unfamiliar people, and preparation for being alone, prevents these problems from developing during adolescence.

Very young puppies will naturally approach unfamiliar people, other dogs, and other animals, and they readily dive into unfamiliar situations. However, as they grow older, the tendency to investigate and approach decreases, as the tendency to avoid the unfamiliar increases. Becoming wary of "the unfamiliar" is an evolutionary adaptive strategy that has obvious survival value for young canids in the wild. Canid pups and cubs naturally become socialized to their own kind and especially their family around the den, and as they grow older, they naturally avoid *unfamiliar* animals that they encounter farther from the den.

Domesticated dogs have inherited this same developmental programming. Although we cannot undo this, we can raise very young puppies to be comfortable with unfamiliar people, places, dogs, stimuli, and situations, so they don't overreact during adolescence.

Adolescent anxiety and stress are *excruciating*, and they trash your dog's quality of life and yours. But they are predictable and falling-off-a-log easy to prevent.

Recognizing and Addressing Anxiety

How many times have I heard an owner say, "But he's never done that before." Or, "She was absolutely fine until ... [insert reason here]." Well, he just did that and she acted that way: She barked and growled at visitors, snapped at a child, got into her first scrap with other dogs, or freaked out because the wind caused a sunshade to flap. "Where on earth did that come from?" people ask.

Adolescent-onset temperament changes go largely unnoticed at first because puppies are seldom challenged in their very limited, familiar physical and social environment. For months, owners live with a seemingly confident and friendly pup, but that's because their friendly, confident dog has mostly been at home — in a *familiar* setting, interacting with *familiar* people, with *familiar* routines, and maybe one or two *familiar* resident dogs. Thus, people are shocked and utterly dismayed when, seemingly out of nowhere, their five-month-old adolescent startles and freezes at a perfectly innocuous stimulus, such as an umbrella being opened.

Because of the sudden onset, people desperately search for a single cause to explain this unexpected change in temperament: *Surely it must be because of an early traumatic encounter or some precipitating, terrifying event.* However, rather than a single scary event *causing* the problem, it is more likely that the event *reveals* that there is a more serious underlying issue: Their dog lacks confidence due to an earlier impoverished environment and lack of social stimulation. Regardless, whatever the cause, that's history, and worrying about it won't help. Instead, start to address the issue immediately through intensive classical conditioning.

For example, our dear little super-socialized Hugo once had a meltdown when he saw an eighteen-inch stone statue of an elf in Calistoga. To help him muster up the courage to approach that statue, we offered him treat after treat, which he refused. Tentatively, repeatedly, he inched closer and closer, only to retreat, and then he did it again and again, stretching his neck to sniff the little elf. After fifteen minutes, he finally sniffed the elf and that was that. He got over it. We walked back and forth several times, and he showed no elf-interest whatsoever, but instead he readily accepted many food rewards.

Dealing with fear is urgent and important; otherwise it will fester and consume a dog's brain. The younger the puppy, the quicker the resolution. Classical conditioning is all about acclimation. For instance, if your dog is afraid of people, dogs, and the great outdoors, every day go to a park and sit on a bench in a quiet setting. For a few hours, just sit and allow your dog to "hide and peek" and watch the world go by. Whenever there's any change in the environment, hand-feed your dog kibble, piece by piece (let this be your dog's meal). If passersby want to greet your dog, explain that the dog is really scared and ask whether they could take the time to toss a few treats (that you provide). As a surprise for your dog, arrange for a few family members, friends, and human members of your dog's core social group to walk by and say hello.

This technique is much more effective than walking through an ever-changing environment with zillions of stimuli constantly bombarding your dog's brain. It is difficult to classically condition a dog in motion because the scary stimuli come one after the other or several at a time. It's too much, too quickly. Instead, first settle in a quiet area with few people and little traffic, then find a bench in a busier area, and eventually sit in an outdoor restaurant on Main Street, or *outside* of a dog park. Build to this gradually; with an adult dog, this is an every-day, several-week process. But your dog will be so grateful that you took the time to help them overcome their fears.

Social Savvy: Dogs Learn from Play

Off-leash play is essential for dogs to develop bite inhibition, impulse control, good manners, and social savvy. Play offers dogs a fun opportunity to learn, practice, and fine-tune all aspects of their behavioral repertoire. They learn the rules of dog society and master getting along with other dogs, including dogs that may have less social savvy.

From all the posturing and possession games, from the squabbles and scraps, young puppies and dogs solidify dyadic and triadic relationships. When dogs learn how to settle most minor disputes during early development, there is seldom any reason for major disagreements or physical fights as adults. After oodles of scraps and maybe occasional fights during puppyhood, the established social rank becomes the *status quo*, which serves to prevent squabbles because the outcomes have been predecided.

Play accelerates and fine-tunes social decision-making skills. In fact, if puppies are given the opportunity to develop social savvy, then as adults, when they meet unfamiliar dogs, they can make massive, all-important decisions in a flash and without bluster or kerfuffle, such as "who has prime access to this meaty bone?" Puppies are renowned for making loads of social blunders, which is the way they learn not to make the same blunders as adults.

Atmosphere Cues

Play teaches dogs to advertise their friendly intentions prior to play with frequent proactive atmosphere cues, which signal that all behaviors that follow are friendly and playful, and not aggressive. Dogs frequently pause play and resignal friendliness numerous times before reengaging. As a guide when assessing an unfamiliar dog, I count the number of friendly and appeasing atmosphere cues per minute. I call this the "friendly quotient." Most friendly dogs average thirty to forty per minute. Labs and pit bulls usually put out over sixty friendly signals per minute! For comparison, watch out for dogs that exhibit zero friendly behaviors but instead stand rigid with only their eyeballs moving, or sometimes their eyeballs are stationary.

Characteristically, friendly atmosphere cues are numerous and fast moving. They comprise: bending the elbows, such as raising a front paw, pawing, playbows, and rolling over; showing the tongue, such as "lizard tongue" (flicking the tip of the tongue) or Labrador-lolling tongue (a protruding tongue is antithetical to biting); and weight-shifting movements, such as quickly and rhythmically shifting weight between front legs, butt wagging, rapid jerky movements, mad scampering, and a jaunty rocking-horse gait.

Once at an Association of Professional Dog Trainers conference, I illustrated the importance of atmosphere cues during dog play by pretending to have a meltdown. At the start of my lecture, I calmly asked for my video to start and waited patiently. Then I asked again, then again. Then I pretended to get annoyed, "Look, I can't start my presentation without the video. Can we have the video *please*!" When the video still didn't start, I escalated my reaction, "Oh this is *ridiculous*!" I asked the panel moderator, "Can't you do anything about this?" Ahead of time, I had told the A/V people what I was doing, but not the audience, the rest of the panel, or the moderator, who froze like a rabbit caught in headlights. "This is POINTLESS! I'm through. *Pathetic!*" I ripped up my notes, threw them on the floor, and stormed off the stage. Everyone was dumbstruck and silent.

As they started to murmur, I happily returned to the stage and said, "That's the importance of atmosphere cues. I have no video, as the A/V people know. Let's restart my lecture, but this time, you know that whatever happens next is not for real. It's a joke. I'm just pretending." Then I groaned, grabbed my chest, and fell to the floor. After a brief shock, people started laughing as I crawled around the lectern on all fours and lifted a leg. Behavior can be alarming when intentions are misinterpreted.

Dogs signal their friendly intentions over and over, especially when encountering unfamiliar dogs. After mutual olfactory investigations, play doesn't start until both dogs have repeatedly signaled that they pose no threat but are friendly and want to play. Sometimes older dogs signal puppies to build their confidence, and other times puppies signal older dogs to get them to reaffirm that they are indeed "just playing."

Adult Dogs Teach Puppies

Another reason to develop your dog's core social group through regular playdates is to provide opportunities for older dogs to mentor puppies and adolescents. Just like people, dogs learn social savvy from their peers and from older generations.

One of the best teachers I've ever known was Big Red Claude, who mentored many dogs in need of social savvy, both during his stay at the SPCA and with us. One vivid example was Pistol, a ten-week-old Malinois puppy that was barely the size of Claude's head, but *he was a pistol.* When they played, both snarled and growled like it was a cage fight, and onlookers were often perturbed. So I filmed one of their "duels" and played it back without sound. Vocalizations, especially barking and growling, can blind us so that we no longer see or read a dog's body language. Without sound, though, the sheer number of friendly behaviors became immediately apparent — an outstanding fifty or so per minute from both dogs. Another amazing thing also became obvious: Pistol controlled the tempo. Every time Pistol disengaged or went still, Claude stopped playing.

Since then, I have paid attention to which dog controls the tempo of play sessions and calls for time-outs. When evenly matched, this can be either dog, but it is not unusual for the initiation, intensity, and breaks in play to be determined by the puppy or underdog. I have blatantly plagiarized many of Claude's teaching strategies for teaching over-the-top, hyperactive dogs to calm down.

I fondly remember a lovely little pit bull in the Berkeley shelter that Kelly and I decided to foster over Christmas and New Year's. I renamed him Flash because his activity was expressed at levels unmeasurable by any scientific instrument. I tried to film him, but he was too fast for me to keep him in the frame, except when, periodically, he would go splat on the ground, front legs wide apart, with a happy, open-gaped, maniacal expression, and then he'd explode in a whirling blur once more. We loved Flash but his hyperactivity was extraordinary.

One day I was trying to teach Flash impulse control on the lower (dog) lawn when Big Red Claude sauntered down to sniff. Flash's behavior changed in an instant. He was now Mr. Perfection. Calmness personified. He offered Claude sequential Sit-Stays, Down-Stays, and Stand-Stays while staring into space. I simply didn't believe it, but luckily, I filmed it. Claude didn't bark, growl, stare, or do anything out of the ordinary.

Later, though, reviewing the video, I could see what was happening. When Claude wanted attention and an appropriate reaction from Flash, he would stop his lazy meandering and intent nose-vacuuming of the lawn and stand perfectly motionless. He didn't move a muscle or even look in Flash's direction. He just raised his nose from the grass and froze. Flash's reaction was like an electric shock — a lightning-fast Sit or Down, and then he was a rock-solid statue, gazing at nothing. Once Flash had a grip on himself, Claude would turn, amble toward Flash, give him a little nose-prod, and then raise a paw to signal play. Reticent at first, Flash would engage. The whole sequence became a daily routine. Flash would patiently Sit-Stay or Down-Stay while watching Claude complete his sniffing duties, which could easily stretch to half an hour, and then Claude would raise his paw and the games would begin.

The Fourteen Most-Common Subliminal Bite Triggers

Dogs seldom bite for a single reason. Instead, the presumed precipitating factor is usually the *last of many* subliminal bite triggers, which eventually exceeds the bite threshold. Dogs become increasingly on edge when several subliminal triggers occur together, at which point a single final straw, even a minor irritation, can put the dog over the top and they react. However, if the dog developed bite inhibition during puppyhood, there will be no injury.

By far, the most-common bite trigger for a dog is when *a family member grabs the collar*.

1. Grabbing the collar, touching the neck
2. Touching the ears
3. Touching the muzzle
4. Touching the paws
5. Touching the goolies or nether regions
6. Hugging or restraining
7. Making a kissy-face or staring
8. Touching the food bowl, bones, or other valued objects
9. Children approaching
10. Men approaching

11. Strangers approaching
12. People acting weirdly
13. Superstitious stimuli
14. Unsettling environments

Case history: A dog lived with a six-year-old boy, who invited his best friend and family to visit. The visiting boy toddled over to retrieve a toy next to the dog, which was chewing a bone, and accidentally brushed the dog's collar. The owners stated that "the dog had always been perfectly trustworthy but bit the child suddenly, without any apparent warning or reason." However, after further questioning, the owners revealed that they lived quiet lives, and they euphemistically acknowledged that the dog "took a while to warm to strangers," was not "overly fond of children," was "a bit dodgy around bones," and was "a little hand-shy." There we have it, the presence of seven or more triggers: (1) male, (2) child, (3) stranger, (4) acting weirdly, (5) bone, (6) collar, and (7) unsettling environment with the clamor of visitors in the house. Also, maybe the dog was just having a "bad day," or felt tired or sick. The dog had multiple reasons to feel anxious and stressed, and the dog most certainly warned the little boy prior to biting, but the warning was likely subtle and quick. Further, the dog most certainly had been warning the owners for years, but the warnings went unnoticed or unheeded.

Please notice that thirteen of the above subliminal bite triggers (all but superstitious stimuli) can be desensitized, easily and quickly, by very basic handling exercises and socialization during puppyhood; in fact, before six weeks of age. Prevention is just too easy. Moreover, the multiple subliminal bite trigger model facilitates resolution by desensitizing each subliminal trigger one at a time, safely below threshold.

To paraphrase Frederick Douglass: It's easier to build strong puppies than to repair broken dogs.

Chapter 3

Five Reward Training Techniques

There are five basic reward-based training techniques. To give you a feel for how they differ, this chapter evaluates their pros and cons in terms of purpose, ease, speed, effectiveness, and enjoyment:

1. Classical conditioning
2. Autoshaping
3. Wait-and-reward training
4. Shaping
5. Lure-reward training

By their very nature, all reward-based training techniques are enjoyable for both dogs and people, and enjoyment always increases substantially along with effectiveness and success. I'm a firm believer that we should use what works best in each situation. What's important isn't fidelity to a single philosophy but recommending whichever techniques are easiest, quickest, and most effective for dog owners to use in each situation.

Classical Conditioning

My grandfather always used to enlist the help of grandchildren plus the village kids to teach one-day-old calves to drink from a bucket. I remember those days fondly. It was a fiasco with most children daubed in splashed milk formula, and young calves happily licking us with their raspy tongues. Occasionally, a bull calf accidentally found his way into the group, and if you bent down, you got butted and flattened.

Years later as a veterinarian and behaviorist, I asked Gramp if it wouldn't have been easier and quicker to let the farmhands teach the calves to drink. His, as usual, sage reply: "As adults, the calves are going meet lots of hikers walking along the public footpaths through the farm, and so I wanted the calves' first impression of humans, in this case little humans covered in milk formula, to be a pleasant one." Children mean milk. Amazingly prescient. My grandfather left school at twelve to work the farm, yet he essentially understood and appreciated the importance of classical conditioning, associative learning, and early neurological stimulation, although he had never heard of the terms. All the animals on the farm were handled and gentled as youngsters mainly by children, not just the puppies and kittens, but also the calves, foals, piglets, lambs, and even the chicks. Early handling was always first and foremost.

Associative learning has existed forever — long before it was "discovered" and given a formal name and scientific definition. It's how animals learn the predictive value of stimuli in the environment. For example, the smell of a rabbit means potential dinner at the end of the trail, or a sudden noise or peripheral movement may be a sign of potential danger.

Pavlov and his bell? In Pavlov's laboratory, the sound of a bell — well, actually, it was a metronome — was a neutral stimulus that had no relevance for dogs and no effect on their behavior. However, after the metronome was followed by meat powder on several occasions, the dogs learned that the sound of the metronome *predicted* the smell and taste of meat, and so they began to salivate whenever they heard the metronome.

We observe similar associative learning with our dogs at home. How many chimes of a doorbell does it take before our dogs associate the doorbell with *visitors*? *Yay!* The sound of a doorbell represents a change in the routine, oodles of attention and affection, and maybe *treats!* It becomes a reason for dogs celebrate and, in some households, give voice and jump with joy.

Similarly, after a couple of mealtimes, the dog is right there when you put down their dish. The next day, the dog is there when you open the feed bin. The next day, the dog is there when you enter the kitchen, and the next day, the dog is already sitting and waiting patiently next to the cupboard where you keep the food bin. And should dinner be later than usual, your dog will let you know — your dog speaks, and you obey. Dogs easily learn lengthy cascades of predictive cues that eventually lead to some pleasant or unpleasant scenario.

During the long course of domestication, the precursors to modern dogs gradually changed their views about us. In the wild, humans were probably the scariest stimulus on the planet. But gradually, the wild species that became dogs and cats and farm animals associated humans with a steady, year-round supply of regular meals, fresh water, and yes, interspecies companionship.

Purpose

All reward-based training is based on the same principles as classical conditioning and associative learning. The owner predicts food. Training predicts food, and praise after dogs sit on cue often predicts food. Then, children, men, visitors, and strangers predict food, and seeing other dogs predicts food and a frolicking good time.

The goal of classical conditioning is to change dogs' perceptions by teaching them to have good vibes and feel confident around every potentially scary aspect of the environment, especially people, other dogs, and unfamiliar scenarios. Classical conditioning is essential for desensitizing predictable subliminal bite triggers when handling young puppies.

Ease

Classical conditioning is the simplest training technique in existence. Classical conditioning is the first exercise taught to volunteers in Open Paw's shelter behavior and training program. Each volunteer is instructed to complete three laps around the shelter, and as they walk up to each cage, toss a couple of pieces of kibble. That's it. In just a couple of days with fifty volunteers walking their rounds each day, the dogs eagerly look forward to people approaching their cages. That simple.

Speed

Classical conditioning works surprisingly quickly with young puppies, causing huge beneficial effects on feelings, confidence, and consequential changes in behavior. However, confidence-building takes progressively longer the older the puppy. For example, whereas imbuing confidence in a shy or fearful eight-to-twelve-week-old pup can usually be accomplished within the hour, rehabbing a five-month-old fearful adolescent may take a month or two, and an eight-month-old dog adolescent usually requires several months. Giving the gift of confidence to fearful adult dogs is a commitment, but believe you me, it's worth it for the dog's sake.

Classical conditioning, socialization, and handling during early puppyhood are the secret to success — in the breeding kennel, in the puppy's new home, in the veterinary clinic, and on walks in a variety of places. In addition, classical conditioning never stops. Think of it as a never-ending but enjoyable way of living with your dog. Living in our world, your dog can never be *too* confident.

As a ritual, whenever one of our dog's died, the following day, I used to place their final classical conditioning treats on their gravestone and say a few teary-eyed words of gratitude, reflecting on their braveness, trust, and sweetness throughout their all-too-short lifespan. "Sleep well." Then I would instruct the other dogs, all in Stand-Stays, "Free dogs," and they would eat the treats. The king is dead; long live the king.

Effectiveness

Classical conditioning seldom fails. Because it's so darn simple, it's very difficult to screw up. Unfortunately, because of the simplicity, a lot of people don't take classical conditioning seriously and don't bother. They assume that a confident dog in familiar surroundings will be confident in all situations, which of course is not true.

I was once watching obedience trials at the San Francisco Cow Palace. An extremely handsome and well-trained German shepherd entered the indoor arena and immediately hit the deck and cried, howled, and screamed as if being beaten. The dog had never been in such a large, noisy, and echo-prone auditorium before. The dog's confidence and training evaporated the instant the dog stepped over the threshold. When something like this happens, we simply cannot let a dog live with the problem; immediate rehab is beyond urgent and requires multiple, repetitive entries and exits into similar buildings. But especially, it means taking time inside the building to slowly allow the dog to become accustomed with the surroundings.

Enjoyment

There's little not to like about classical conditioning techniques. Look at the calmer and quieter shelter dogs. Just observe the smiles on people's faces and the dogs' wiggles and wags as they sit to greet visitors or meet strangers. Just think how good you feel when your dog sits and stays and gazes up at you. Marvel at your dog's confident demeanor when meeting unfamiliar dogs. Pat yourself on the back while watching your dog obviously enjoying calm interactions with children. Be amazed that your dog is not reacting to the general hubbub and cacophony of downtown construction, the horrendous traffic, and the Brownian motion of the throngs of unfamiliar people. All because of early enrichment and classical conditioning.

Autoshaping

In the early 1900s, the concept of autoshaping was developed during laboratory studies on animal learning. Researchers designed and built special training cages that provided various types of consequential feedback, primarily food pellets and/or shocks, as well as cues or warnings for the impending consequences. The researchers would put the animal in the cage and come back later to collect and analyze the data and publish the results.

Although not directly related to people training dogs, these devices trained animals without the need for people to be present. With autoshaping, basically, animals train themselves.

For years, the big question in dog training was: How can I punish my dog for barking, chewing, housesoiling, and so on, when I'm not at home? Not too long after Kongs came to market, I had an unexpected epiphany that made me realize this was the wrong question. I was giving a seminar, and a gentleman in the front row kept asking about his Doberman's barking, chewing, and hyperactivity. I suggested giving the Doberman a Kong to chew.

"Tried that. My dog doesn't like it!" he said.

So I kept giving him suggestion after suggestion to increase the dog's interest in Kongs: Waggle the Kong, tie a string to the Kong and play tug, play fetch with the Kong, play find the Kong. After every suggestion, he retorted, "Done that, doesn't work!"

I must admit I was getting a little exasperated, and I just blurted out, "Then ... stuff the Kong with food!" The idea just popped out of my brain, but as I said the words, I thought, *Wow! That is one really great idea! Reward-only autoshaping.*

To this day, the notion of feeding dogs from food stuffed in hollow chewtoys is one of the best ideas I've ever had. Just a couple of days of feeding from Kongs absolutely reprograms a dog's brain: They bark, chew, and pace 90 percent less because so much of their day is spent lying down chewing their chewtoy and being *repeatedly rewarded* for being calm and quiet and chewing a chewtoy by each little piece of food they extricate.

Indeed, the question we should be asking is the same as always: *If this is wrong, what is right?* In this case: How can I *reward* my dog when they're out of reach or home alone? Obviously, by giving them food-stuffed chewtoys.

When I got home, I started feeding Omaha only from stuffed long bones and Kongs. The changes in his behavior were immediate and colossal. I tied the stuffed chewtoys to the leg of my chair, and he would lie down next to me and chew until he fell asleep. Spending more time next to me, he changed from an independent soul to a loving soul that thrived on companionship.

After a couple of weeks, Omaha would follow me around the house, voluntarily lie down next to me, and occasionally rest his chin on my knee. We did everything together. We jogged thirty-two miles a week on the hill trails, and when competing in obedience, we would camp in a tiny pup tent at the showground. At night, Omaha would trot into the tent and lie down in a straight line at the far end and be my pillow.

Purpose

Reward-only autoshaping using Kongs teaches dogs to "teach themselves" to be calmer, quieter, and to chew chewtoys.

Ease

It's difficult to get much simpler than letting dogs train themselves. All you have to do is regularly clean and restuff the chewtoys, and your dog will self-train for hours on end.

Speed

Time is required to stuff hollow chewtoys with food, but thereafter, *it's the dog's time investment, not yours*. The more time your dog spends chewing stuffed chewtoys, the greater the autoshaping effect. Just give your dog a stuffed chewtoy on their bed or in their crate, and watch the magic happen. Your dog will soon settle down, calmly and quietly, preoccupied by trying to get the food out of the chewtoy.

To reduce your chewtoy-stuffing time, stuff several at the same time. After Kelly developed the Open Paw program, shelter dogs were fed only from Kongs, and on Friday evenings, we would meet up with our volunteers at a pizza parlor and stuff several hundred at once. A bakery donated cardboard trays to transport our stuffed Kongs and store them in the freezer.

Effectiveness

Autoshaping is *extraordinarily* effective, especially when dogs are fed only from hollow chewtoys for a couple of days. Use some kibble from your dog's daily ration as food lures and rewards for training and feed the remainder in chewtoys. In other words, suspend feeding from a dog bowl until you're satisfied with your dog's change in behavior, then feed your dog how you prefer. Every little piece of food your dog extricates from the Kong rewards your dog for what they are doing — *lying down and quietly chewing a chewtoy*, rather than running about, barking, or trashing the house.

The longer your dog spends chewing chewtoys, the longer and more frequently your dog will be rewarded for lying down, so learn how to stuff chewtoys to prolong food removal. Initially, to get the dog interested, use dry kibble, which comes out easily. Then use moistened kibble, and then stuff the Kongs with kibble mush and freeze overnight. Voilà! Kongsicles!

There is a wealth of information about prolonging food removal from chewtoys to lengthen the autoshaping effect. For example, smear honey inside, which is sticky, plus it has antibacterial properties. Cap the top with peanut butter, although this is often overdone and adds too many calories. Insert a few rocky-road niblets of freeze-dried liver or bacon bits in the bottom of the Kong prior to stuffing with kibble mush. Push a dry biscuit in the top of a kibble-mush-stuffed Kong, which acts as a partial plug.

Since dogs are crepuscular, and early morning and late afternoon are the peak periods for activity — including undesired activities such as excessive barking, chewing, and hyperactivity — the two best times to offer your dog a stuffed chewtoy (or two) are right before you leave for work in the morning and right after you return home.

Feeding from chewtoys has many additional benefits. For example, predicting the need to eliminate. After a lengthy chew, most dogs fall asleep, and as the dog sleeps, their bladder slowly fills. So after an hour, wake your dog, show them the appropriate toilet area, and teach the dog to eliminate on cue.

The longer dogs chew Kongs to get to the food, the quicker they become Kongaholics, and the longer they will continue to chew Kongs even when they're empty. Well, we think they're empty, but a dog's nose indicates otherwise. Yes, there's always a tiny morsel or an odor trace left inside.

Most important, puppies that are Kong-fed, and that have frequent solitary downtimes enjoying their Kong habit on their own, seldom develop separation anxiety as adolescents.

Enjoyment

I could watch a dog chewing for hours. And I often do. Dogs seem so content and involved when engaging their chewtoy hobby. It's calming, relaxing, and entertaining for all to watch, especially other dogs.

Wait-and-Reward Training

I simply cannot remember my dad ever shouting at me. The only time I can remember my dad shouting was when our generator refused to start, and it was getting dark, and we had no lights. My dad had some unique ways to teach us to behave when we were kids. When my sister and I sat in the back of the car, we couldn't resist annoying each other. I used to check that Dad was focused on the road and then poke her in the arm. She would shout, "Stop it!" and poke me back harder. This continued until Dad's patience wore thin. Normally, Dad lived in his own special world, lost in thought. However, certain triggers brought him back to earth: bad workmanship, extreme stupidity, certain politicians, and unruly children. My dad's response to child noise and silliness in the car was to casually check the rearview mirror, slowly pull the car over to the side of the road, park, apply the hand brake, turn off the ignition, light his pipe, and open the newspaper. All this without saying a word.

After a while, we would notice the car was no longer moving, and say, "Come on, Dad, we'll be late for school!" But he just sat there, reading, and puffing away. Periodically, he looked up to glance at us in the rearview mirror, but he didn't budge until both of us had stopped talking and were sitting perfectly still, our arms folded — whereupon he lazily folded the newspaper, put his lit pipe in his tweed jacket pocket, started the car, checked the mirrors, and proceeded. I really have no idea how "folded arms" became part of our silent apology and reversion to acceptable behavior. I guess, when you know something is up, but your dad is unresponsive, you try a whole bunch of things to return to family life as normal. From then on, if ever Dad glanced at us in the rearview mirror, the two of us ceased talking and sat up straight with our arms folded. It became a joke and made him smile. But then, he so often had a hard time keeping a straight face.

That is the essence of wait-and-reward training: Just observe the dog, *wait* until they do something you like, and then *reward.*

As part of Open Paw's shelter behavior and training program, after taking three laps around the cages classically conditioning the dogs, the volunteers take two more laps, but spend a little time outside each cage observing the dog's behavior and waiting for one of seven behaviors (listed below) to suddenly appear, and then they praise and offer kibble.

1. Attention — glancing, gazing, or looking at the volunteer
2. Approach, or remaining in proximity
3. Sit, or Sit-Stay, or the act of sitting
4. Down, or Down-Stay, or the act of lying down
5. Stopping bouncing and having "four on the floor"
6. Stopping barking or not barking
7. Looking cute

All these behaviors are all-or-none, either/or situations: either the dog is not looking at you, not approaching, not sitting, or not lying down, or they are. Either the dog is barking and bouncing, or they're not. The transition between the two is abrupt and complete. There is no need to take a long time building a desired behavior gradually and progressively (as with shaping); when the behavior comes, it appears *all at once*.

Within just a few days, the shelter becomes a calm and virtually *barkless* facility. When any person approaches a cage, dogs automatically sit, shush, and watch the person. Simply amazing. The relief for dogs, staff, and volunteers is enormous.

Why is "looking cute" on the list? Because in dog shelters, we've found that this is the most effective way to increase the likelihood that a dog will be adopted.

Purpose

Wait-and-reward training is tailor-made for teaching out-of-control, over-the-top, inattentive dogs to develop impulse control and to learn to chill, engage, focus, and pay attention, without the trainer giving any instructions whatsoever, and without using a single reprimand or punishment. Instead, you wait for what you want, and then the moment desirable behaviors appear, you praise and reward the dog.

Wait-and-reward training is like waving a wand when teaching Off, Settle, and Shush. Also, the technique is mind-bogglingly brilliant for teaching dogs to walk calmly on-leash and to sit by our side automatically whenever we stop. The only words we need to say are "good dog."

Ease

Wait-and-reward training is the second-easiest training technique on the planet, next to classical conditioning.

Speed

Initially, patience is required while waiting for the first glance, the first quiet moment, or the first sit. Thereafter, things speed up. For example, when leash walking, the second sit comes quicker, the third quicker still, and soon, automatic sits are coming fast and furious.

That said, some dogs take a while to "get it." If that happens, just watch their antics with amusement. Sometimes, though, I comment on the dog's misbehavior and follow Will Ferrell's advice by responding with scathing sarcasm if the dog continues acting like a loon for long: "Do you reeeally think that's worth a piece of kibble?" "A sloth could learn quicker." "You know, you bounce like a cowpat." "Did your mother not tell you that you have the bark of a gerbil?"

Sometimes my feedback speeds things up; talking captures the dog's attention and they still themself to listen. Or maybe saying anything makes the dog feel acknowledged. Who knows? But this is how I amuse myself when waiting for specific desirable behaviors.

The longest I have ever had to wait for a dog to sit was twenty-two minutes — a dog trainer's dog in LA — I think it was a Weimaraner. The second longest was eleven minutes, and it happened to be in front of a large workshop audience in Belgium, comprised almost entirely of experienced male trainers with their Malinois and Belgian shepherds. The only companion dog was an adolescent Labrador that was beyond out of control. The poor owner was feeling ostracized and unwelcome, but I told him his dog was perfect for a demonstration.

The audience appeared doubtful from the outset, and so I thought, *Don't bother lecturing, show them*. After introducing myself and thanking them for coming, I took the Labrador's leash, and he dragged me to the center of the room, where I stood my ground and waited for a sit. Larry, as I called him, since I couldn't pronounce his Flemish name, became seriously deranged, even by enthusiastic Labradorian standards! He ran round me three times, hog-tying me with the leash, wagged his tail and butt furiously, barked, jumped, hooked a paw in the breast pocket of my jacket and ripped it off, backed up, and barked some more. Then he realized I had treats on my person and it got worse. He started to chew on the side pocket of my beige lecture suit, turning it a disturbing shade of brown. He had me down on the floor twice, and all this while I was waiting for the first sit.

Larry's hyper behavior eventually slowed down, but he was still barking nonstop. If I have just one stellar quality as a dog trainer, I never give up. I went into my "calm zone" and slowed my breathing, and then ... all of a sudden, quite out of nowhere, after eleven minutes to the second, Larry abruptly stopped barking and plopped himself into a sit. Praise, praise, praise, six treats in succession. Then I took one giant step and stood still to wait for the second sit. The instant I moved, Larry erupted once more, but only for two minutes. Second sit. "Good boy, Larry, goooood boy," and I gave him three treats. After one step and thirty seconds, a third sit, and after ten seconds, a fourth, and we were off and running.

After Larry gave me half a dozen immediate, automatic sits, I started taking more steps before stopping and waiting for a sit: two steps, three, four, six, eight, ten, fifteen. Then we walked round the room with Larry on a loose leash, fast straight-line heeling, numerous ninety-degree and about-turns, figure eights, and changing pace. Then I started to give a running commentary to the audience while on the move: "I'm going to speed up now to improve his gait and straighten him out.... I'm going to slow down now to get him to slow down and look up at me." And Larry slowed down and looked up at me, sending thoughts with his eyes, *This is cool!*

The only words I said were *good*, *boy*, and *Larry*, and Larry didn't even speak English. Yes, there was an enormous eleven-minute wait until the first sit, but after another three minutes I was giving my running commentary to explain what they were seeing. Belgian Larry starred in my all-time favorite demo for so many reasons: The audience was comprised of disbelievers (but they came, they were there), the transition was so enormous and quite abrupt, and initially, Larry was soooo good at being "bad." At the outset of the workshop, the audience was just dying to disagree: *A long-haired PhD from California. Puh!* They were all leaning back in their chairs, arms folded, but then ... the unexpectedness of the extreme Hyde-to-Jekyll transformation was impossible to believe. But there it was, right before their eyes.

Effectiveness

This training technique is just bread from heaven for resolving the antics of noisy, rambunctious, and inattentive dogs, whose exuberance is frustrating beyond words. Often, out-of-control behavior is only exacerbated at the sight of food or with any attempt at physical correction. However, by not reacting, and by simply observing and waiting, these crazy behaviors wane, wilt, and wither, and then suddenly, the dog hits on a ploy that works: The dog sits, and we praise and reward. Since rewards cause behavior to be more likely to occur in the future, the sits come quicker and quicker, and last longer and longer, and eventually, they choke out the silly stuff.

Originally, I termed this training technique "all-or-none reward training," since it works best with either/or behaviors. The behavior we want instantly goes from "none to all" with no in between. To be honest, I never really liked the term. Descriptive, yes, but a bit of a mouthful. Years later, Jamie came up with the term "wait-and-reward," which I love and now use all the time — although I don't use the acronym because it's a mite combative. However, Joel Walton, a good friend of mine, a breeder, and owner of the *original* Larry Labrador, came up with a stellar acronym in his delightful little book, *Positive Puppy Training Works*: SOAR, which stands for "simply observe and reward."

Enjoyment

You just reward the dog for any behavior you take a fancy to. One of the beauties of this technique is that you don't give any instructions. Consequently, your dog can never be noncompliant, and so you don't become frustrated. Basically, wait-and-reward is like shaping on steroids, but instead of using a click and a treat, you use praise and kibble.

Maybe the most wonderful aspect about wait-and-reward training is how calming and joyful it is for people. Training older puppies and adolescent dogs can be tiresome and frustrating, especially when they become distracted, ignore repeated commands, and blow off food lures. As soon as an instruction, request, or command is given, we have expectations, and we get upset when they are not met. However, with wait-and-reward, we give no instructions, we have few expectations, and so *our dog can never be "wrong."* That means there's no reason to get upset, reprimand, or punish. We just wait, watch, and reward. Moreover, waiting patiently for the first few desirable responses helps mirror the calm demeanor we want from our dog. Wait-and-reward training puts us in a wonderful mental space.

Indeed, when dogs and their owners emerge from a wait-and-reward adolescent class, they're all happy. People entered the room with an out-of-control beast, and they leave the class with relaxed dogs with lolling tongues, slowly wiggling butts, and waggy tails. Plus, the owners are smiling and laughing, looking relieved, relaxed, and refreshed. They are proud of their dogs, and most likely their dogs are proud of their owners' patience and calm. Yes, patience is a virtue and is so soothing for the brain.

Shaping

Shaping is the principle behind "clicker training" — breaking down complex behaviors into a series of achievable steps. The trainer "marks" (clicks) and reinforces successive, predetermined approximations toward the final performance in a gradual and progressive fashion.

Whistles and crickets have been used in animal training as "markers" of desired behavior for years. Starting in the 1930s, Keller and Marian Breland and Bob Bailey pioneered the art and applied the science of shaping while training military dogs (and cats) and later a variety of species for training shows at county fairs. As with all the techniques I've mentioned, shaping is so old that it's new again.

In the 1990s, the first dog training workshop using clickers was hosted by Kathleen Chin of Puppyworks. Karen Pryor lectured, and Gary Wilkes demonstrated with his inimitable Australian cattle dog. Initially, the seminar was under-enrolled because many dog trainers didn't get it. How could a "clicker" possibly help to train dogs?

Of course, the clicker was not the revelation, reintroducing *shaping* was the biggie, especially given the climate of dog training at the time. Any convenient marker would suffice: a beeper, the whistles of old, snapping fingers, verbally clicking, saying "yes," or my choice, "G'dog." My mouth has always been quicker than my thumb.

At the time, I was already familiar with shaping from my days at Berkeley, and I used the Brelands' *Animal Behavior* book as my choice of a text when I was a teaching assistant in Dr. Beach's animal behavior course. I thought shaping offered something very special for dog training: a nonaversive means to teach nonretrieving dogs to retrieve and have fun doing it.

I helped Kathleen promote the seminar, billing it as a "secret way to teach a pain-free retrieve." Puppyworks ended up with approximately two hundred people at the workshop. Shaping was back in the fold. Thereafter, Karen Pryor single-handedly promoted and popularized clicker training worldwide, sparking a clicker training revolution.

Purpose

Shaping is a marvelous technique for refining behaviors, and for teaching behaviors that are difficult to lure, which in the eighties, amazingly, still included teaching nonretrieving, working dogs and hunting *retrievers* to retrieve on cue.

Shaping excels at fine-tuning the quality and precision of behaviors, such as heeling, and for capturing cute or unusual facial expressions, such as single or double flew-puffs, an open-gape "smile," "soft eyes," "grinning," and "laughing," or unusual body positions, such as teaching nonpointing breeds "the Heisman."

Shaping also makes it possible to teach behaviors that are not in a dog's natural behavioral repertoire and so are difficult to lure, such as a faux pee (a fake leg lift) or a pirouette on the front legs. Why cue these behaviors, you might ask? Because with shaping, you can.

That said, both these behaviors *are* in the dog's behavioral repertoire. Some marking-crazy male dogs will lift their hind leg over and over, long after they have no urine in the tank. And handstands occur more commonly than you might think. Most people know that dholes (a wild dog from India) and African wild dogs (Cape hunting dogs) routinely urinate standing on their front legs, but so do a lot of beagles. Some beagles both poop and pee when standing, or walking, on their front legs. Maybe to pee high on the tree? An insecurity complex?

Most important, shaping works with dogs that have no inclination to retrieve whatsoever. Some dogs, for instance, have an understandably commonsense perspective to an owner who throws away an object and then asks for the dog to bring it back: *If you really want it that much, why on earth did you throw it away?* These are not bad dogs; they just don't see the point of fetching. They simply have never been taught how happy their person would be if they brought back the discarded object. For me, the gift of shaping will always be that it offered a nonaversive alternative to teach Fetch in the competition, working, and hunting worlds, which once routinely used a "forced retrieve" (which I discuss in chapter 14).

Ease

Clicker training is easy in the sense that it doesn't take much brain power or thumb strength to click a clicker, but it does require a sizable knowledge base of animal learning theory to know *when* to click. Both vigilant observation and ultraprecise timing are essential to click effectively. Shaping requires the greatest skillset of any reward-training technique. I feel that clicker training is like the game of Go — easy to start, but takes an age to master. I've watched Bob Bailey training at length, but I think few will ever achieve that level of mastery.

Speed

By its very nature, clicker training is a slow, gradual, and involved process in order to mark, reinforce, and capture each step of the process. A lure-shaping hybrid technique accelerates training. For companion dogs, shaping is a two-step process: first to "capture" the behavior, and then to achieve reliable *verbal* control. Training with a clicker and treats takes time, and then more time is required to put the behaviors on cue, and then to phase out the necessity for the clicker and bait bag in real-life scenarios, such as at home or on walks.

Effectiveness

The effectiveness of shaping varies greatly with the skillset of the trainer. Consequently, shaping is the best exercise for people to develop the requisite skillset, that is, consistency, observation, and split-second timing. Attending clicker classes will improve anyone's overall training skills no matter which techniques they prefer to use.

However, most basic manners don't require shaping. Come, Sit, Down, Stand, Rollover, Beg, Bow, Back up, Go to Tanya, Go Pee, Shush, Find chewtoy, and so on, may all be quickly lured, or fairly quickly "captured" by wait-and-reward training.

I find a shaping process extremely effective for gradually and progressively increasing the duration of eye contact (Watch), Shush, Sit-, Down-, and Stand-Stays, and when teaching Off. Baby steps offer early and convincing proof that a seemingly out-of-control dog is learning.

For me, though, shaping's irreplaceable gift to dog training is a nonaversive means for teaching Fetch. Nina Bondarenko, an astoundingly creative television dog trainer, rock-drumming motorcyclist, and Rottweiler breeder from Australia, demonstrated this when she co-hosted my UK television program for a season. In under fifteen minutes, by using a lure-shaping hybrid, she taught an unfamiliar German shepherd to retrieve a bunch of keys that she had thrown into the long grass. She dropped the keys, *clink*, and then clicked and gave the dog a treat the moment the dog glanced toward the clinking keys. Then, each time she dropped the keys, she progressively upped the ante before clicking and treating. She waited for the dog to look for longer, even longer, then look and take a single step toward the keys, then two steps, then multiple steps, then to nose-prod the keys, then pick them up for a second, then two seconds, then hold for three seconds, then hold and carry, then hold carry and deliver to hand, and then she threw the keys greater and greater distances, until eventually... into the long grass, wherein the dog plunged and came back with the keys and plopped them in her hand. Of course, finding the keys was simple for the dog's nose. The many baby steps were needed because the complex behavior was taught without instruction.

Enjoyment

What can I say? People love it! Dogs love it! Smiles and tail wags all around, even more so than after wait-and-reward training classes. After clicker classes, it's like dogs and humans are all under the influence of happy pills. Indeed, they probably are. This would be a wonderful study: "The Effect of Clicker Classes on Happy Endocrinology of Humans and Canines."

As with wait-and-reward training, since no instructions are given, the dog can't be "wrong," and this must be such a relief for dogs, not having to listen to owners nigglingly harping on them all the time. All incorrect choices, misbehaviors, and improvisations are simply ignored, so handlers don't become disappointed or dispirited by their dog's performance. Training is never a downer. Dogs just love the party atmosphere and absolutely thrive on their owner's extra and intense attention.

What I really love about clicker training is the camaraderie among trainers within the group. In the dog training world, I've never seen anything like it — apart from some of today's working dog groups and the Association of Professional Dog Trainers' first decade of annual conferences, which were informative and entertaining, oozed togetherness, and were nothing less than riotous, glorious, undiluted fun!

My two biggest criticisms of clicker training are that (1) behaviors are taught *without prior instruction*, that is, without introducing verbal cues from the outset and without luring, and (2) what on earth happened to praise? As well as being a primary reinforcer, praise quickly becomes a very powerful secondary reinforcer. In fact, praise combines the functions of both the "click" and the "treat" by both marking and reinforcing the preceding behavior. I feel that omitting verbal instructions and replacing praise with a click took the voice from dog training. Whereas shaping is the ideal technique for various purposes, and the only technique for some, I feel it made companion dog training unnecessarily complicated, too clinical, and even impersonal.

Lure-Reward Training

Lure-reward training is a fully comprehensive training program that comprises teaching dogs *what* we would like them to do, motivating them to *want* to do what we would like them to do, and producing and quantifying the highest levels of reliability and performance. As such, lure-reward training is always my initial method of choice because of the sheer speed of learning.

Whenever a dog is too distracted to pay attention and engage, I temporarily switch to wait-and-reward training to teach impulse control and focus, then I employ baby steps to increase the duration of their newfound attention and calmness. Then I resume lure-reward training to further expand their vocabulary.

Whereas shaping and wait-and-reward training reduce expectations and pressure by *not* giving instructions, the primary function of lure-reward training from the outset is to teach dogs the meaning of verbal instructions and handsignals, then we test their comprehension by measuring the reliability of responses.

As the name suggests, food (or toy) lures and rewards are used initially as training tools, but then using food is phased out in a stepwise systematic process. Food lures are the first to go, so that the dog's reliability of performance does not become dependent on people having food in their hand, pocket, or bait bag. Once dogs have learned the meaning of handsignals and verbal requests, food lures become unnecessary.

Next, the number of food rewards is drastically reduced. Initially, food rewards are indispensable for people who have difficulty praising convincingly. However, once the dog is getting the hang of things, *reducing* the use of food rewards significantly *increases* their impact, and owners shift to primarily use much more powerful *life rewards*, such as dog-dog play, walks, and interactive games. When using life rewards, reinforcement takes a quantum leap.

However, with other training techniques, it is much more difficult and time-consuming to phase out the necessity of training tools, and in many instances, they become lifetime management tools. For example, when dogs are initially trained on-leash, few develop off-leash reliability, and even on-leash walking becomes a problem, prompting many owners to purchase physical equipment — like specialized collars, halters, and harnesses — to prevent pulling. Similarly, clickers and bait bags often become permanent training accessories.

Lure-reward training is the method of choice for teaching ESL to facilitate better communication, which is an absolute must when instructing dogs *what to do* and *what not to do*. Using *words* in dog training is a game changer, and this is what I explain in chapter 4 and the rest of this book. When we can talk to our dog using words they understand — because we have taught them — our relationship truly blossoms.

Chapter 4

Lure-Reward Training Secrets

When it comes to dog training, people tend to have set views regarding *what* dogs should learn and *why*, without giving much thought to *how* dogs learn. Just as Victorian schoolteachers and parents taught children as if they were small grown-ups, many people approach dog training as if dogs were furry humans who learn like people and have an innate ability to understand language. Then people issue commands and shout as if that will improve a dog's comprehension. Yes, shouting might get a dog's attention, but until dogs know exactly what our words mean, they still will have no idea what they are being asked to do. As a result, dogs are frequently reprimanded and punished for failing to comply or for "misbehaving" when they have no idea that rules exist.

No dog likes to be punished. No dog likes their owner to be frustrated, upset, or angry. Dogs are social animals that yearn to join up and get along with their human pack. However, to do that, dogs need to learn *how* we would like them to act. We must teach dogs *what* we would like them to do and then motivate them, so they *want* to do it. Any dog training method needs to accomplish four objectives:

First: Modify and mold the dog's temperament, especially regarding their feelings and demeanor toward people, so that the dog develops a natural *off-leash* "owner gravity" — a psychological bungee cord and centripetal attraction to family members, which offers proof that the dog loves training and the trainer, that is, you.

Second: Teach dogs the meaning of the words we use for instruction and guidance.

Third: Use a variety of life rewards to motivate dogs to want to comply and to reinforce desirable behaviors so they increase in frequency and squeeze out undesirable behaviors.

Fourth: Use dog-friendly training techniques to resolve any issues with noncompliance and misbehavior, by using words as additional guidance until the dog clearly understands what we want, what's appropriate, and what's not.

Lure-reward training accomplishes all four objectives. Here are some secrets to how it works and why it is the easiest, quickest, and most effective method. But before we start, let's consider our canine trainee, or maybe we should say, our canine trainer.

Dogs Train People Better Than People Train Dogs

Dogs are just so good at training their owners that sometimes it becomes blurry *who trains whom*. We do our little bit of training and teach dogs to Come, Sit, and Settle on cue, but before long, our dog takes over as master trainer.

When preoccupied and concentrating on texts or tax forms, we sometimes get the feeling of being watched and turn to see our dog sitting behind us, their eyes boring into the back of our skull like thought-lasers: *Feed me. Walk me. Scratch my ears.* And we do. It starts with us requesting Sit and the dog responds by sitting, but before we know it, the dog's Sit becomes a request (command/demand) for us to respond with dinner, walks, attention, and affection.

We say, "Fetch," and throw a ball, and the dog responds by running off and bringing back the ball, and then on a sunny Sunday, while we are working through the *New York Times*, the dog's nose pokes under the paper and plops a tennis ball in our lap. The dog requests "throw it" and we do. Once when I was babysitting Lazaretto, Kelly's Belgian commander-in-chief, he plopped a wet, slimy ball in my lap, and so I hid it under the other segments of the paper. He went for it, crumpling the unread sections, and put the ball back where he thought it belonged, in my lap. So, I sat on it, which was wet and uncomfortable. He sat for thirty minutes, riveted on the spot where the ball was last seen. Somehow, one "pain in the arse," that would be Laz, intently focused on my "pain in the arse," took the joy out of reading the paper. Of course, I gave in.

Dogs effectively train us to be a chef, waiter, commissionaire, promenade companion, masseur, chauffeur, porter/caddy (the paraphernalia we carry for dogs), and personal trainer (throw the ball, throw the Frisbee, throw the ball, throw the Frisbee...). And dogs train us to do it immediately! *Chest scratch. Now!* And our commissionaire duties are full time. The sheer number of times in the day we open doors for our dogs, and often on cue, mind you! The dog woofs, or rings a bell on a string, and we get up and open the door for them to go outside, and then to come back inside, to get in the car, to get out of the car, to go into buildings.... It never stops.

Sometimes dogs train us to perform quite complex routines. In one of my puppy classes, I once saw a puppy cocker spaniel teach their owner to go out (walk away), stand-stay, pirouette, recall, and stand-stay once more. All I said to the owner was, "Walk round the room and get your puppy to follow you." The owner strode away, slapping their thigh, and verbally pleading for the pup to follow. The pup walked away and lay down. The owner then stopped, stood still, twirled a full circle to locate the puppy, and then strode back toward the puppy (a perfect recall) and stood stationary with hands on hips.

So many owners perform recalls for their dogs, especially Bernese and mastiffs. The owner calls the dog. The dog doesn't move. The owner promptly returns to the dog and brings out the treats to further reinforce the breed propensity. Even difficult-to-train owners eventually comply. When walking, bulldogs forge and pull to get their owners to accelerate and walk quicker. Beagles lag and sniff to get owners to slow down and wait. And bassets and bloodhounds, masters of the art of training their owners, when done with walking, simply flop down on the ground to train their owners to fetch the car, pick them up, and lay them down carefully so they may stretch out on the back seat.

Another classic is the Chihuahua that boasted to other dogs that he had taught his owners to "speak" on cue. The other dogs didn't believe this and so the Chihuahua told them to watch, listen, and learn. The Chihuahua voiced, *Yap! Yap!* And the owner responded, "Be quiet." *Yap!* "No!" *Yap! Yap!* "Stop it!" *Yap! Yap!* "For Heaven's sake, be quiet!" *Yap! Yap! Yap! Yap! Yap! Yap!*

The other dogs chorused, *Hey! Cool! A duet!*

That's how many of us live with our dogs. But isn't it wonderful? Personally, I love dogs that take control of their daily routine and remind me when it's time to get out of bed, open doors, go for a walk, eat together, couch together, and when the day is done, go to bed together. Isn't this what it's supposed to be about?

Dogs Learn Like Dogs

Teaching a different species is challenging, fascinating, thought-provoking, enjoyable, and ultimately, when successful, fulfilling. Being able to communicate with nonhuman animals is full of awe and wonder.

But it's important to remember that dogs are dogs, and they learn like dogs. Many people approach training as if dogs think and learn like people, or like wolves, and so it is hardly surprising that even the most biddable of dogs can become confused. To compound matters, when dogs learn at a slow pace and fail to grasp exercises quickly, a knee-jerk human reaction is to blame the dog's shortcomings *rather than evaluate our own training techniques and skills.* Typically, many people call a dog stubborn, defiant, dominant, or dumb, or they blame the breed, having just read some pop "science" article comparing breed IQ.

When dogs learn slowly, that might be because their teacher is still learning how to train. In this case, as people become more adept at training, their dog will become a quicker learner. Or the dog might simply be having trouble understanding the task at hand. Or maybe the dog is not being motivated sufficiently to comply.

Just as people learn best in a variety of ways — some are better visual learners, while others learn better by reading or by doing — dogs learn best in their own way. We must adjust our training to what works for dogs, while also adjusting for each individual dog. First and foremost, dogs do not understand our language until we teach them. They naturally respond to tone and volume, but they preferentially respond to movement, especially body language and facial expressions.

Further, whereas humans are huge generalizers (after just one incident of bad service in a Paris restaurant, people might bad-mouth all of Europe), dogs are extremely fine discriminators: They learn *exactly* what we teach, but they have great difficulty generalizing what they have learned to other people and other situations.

For example, if you train your dog in the kitchen at 5 p.m., you'll no doubt produce a wonderful, late-afternoon kitchen dog that responds well to you, prior to dinner, in the kitchen. The big question is: Will your dog automatically generalize that training to other times, places, and people? The answer is — big drumroll, please — probably not. Your dog won't work as well with other family members, that is, until they themselves train the dog. Also, your dog will not work as well with you at other times and in other rooms of the house, in the backyard, and off-leash in a dog park.

Because people often attribute human characteristics to dogs, especially powers of cognition, they frequently assume that dogs understand more than they do. They say, "He knows the command," or "She *knows* it's wrong." But dogs don't necessarily know that what's "right" or "wrong" in a kitchen is also right or wrong in every other scenario. For dogs to learn that, we must train them in every other scenario. This has nothing to do with a dog being dumb or lacking intelligence, whether that's compared to other dogs, species, or people. It's just how dogs learn.

Then again, some people insist that dogs think more like wolves, and by a considerable lupomorphic leap of logic, they try to act like wolves when training dogs. Or at least that is one justification for procedures like the *alpha rollover*, which involves grabbing a dog and forcing them onto their back. First, that's not what wolves do. Second, just as a dog is not a person, a dog is not a wolf, and dogs obviously know that *people aren't wolves*. Dogs may have evolved from wolves, but their behavior and temperament are *very* different, especially because they were selectively bred for *ease* of socialization with humans. Even so, dogs still require early socialization to take full advantage of this inheritance.

Ultimately, we should train dogs in ways that reflect how we would like them to live with *people* (not with wolves or other dogs), using methods that are easy for dogs to understand. For example, we can incorporate one of their languages, body language, to teach the meaning of handsignals, which they learn in a flash. Then we can use handsignals to teach the meaning of our verbal requests, that is, ESL, or English as a second language.

All Breeds Are Equally Smart and Begging for an Education

A major gripe of mine is lists of the relative intelligence of different species and of different dog breeds. Most administer IQ tests that incorporate questions based on human intellect, making them unfair and therefore meaningless. But even when the tests are dog-appropriate and supposedly quantify a dog's cognitive abilities — though more accurately, this is usually their sensory abilities — rank-ordered results strongly affect people's attitudes toward different breeds, and all too often training suffers.

For example, the IQ list has long been topped by border collies, leading many to assume that, if collies are so smart, they don't need to be trained, and so they aren't. Now, if there were ever dogs that need to be trained to be urban and suburban companions, or even rural companions if there are no sheep to tend, it's border collies. The good news is they only need to learn three commands: (1) Settle down, (2) settle down, and (3) settle down!

On the other hand, when people see that Afghan hounds are often listed as the least-intelligent breed, they often assume that these dogs are too "dumb" to be trained, and so they aren't.

All dogs *need* an education. In fact, they are *begging* for an education. No puppy left behind! While there are individual differences in doggy intelligence, all dogs are smart enough to be trained, and none are too smart not to need it.

Another issue is when breeds are rank-ordered in terms of *trainability*. Back in the good old days, trainers did this based on success in obedience trials. The top breeds were invariably those that have been bred to be working dogs. Among the top five breeds were border collies, golden retrievers, German shepherds, and Dobermans, along with papillons as an outlier, while the lowest-ranked breeds included Malamutes, mastiffs, Akitas, Rottweilers, and Afghans.

What bearing, if any, does ranking breeds by trainability mean for your choice of a family companion? Not much. In fact, hardly at all, especially considering that back then rankings were based on dogs trained on-leash, usually with numerous leash corrections. At the beginning of the eighties, off-leash lure-reward training leveled the field considerably and created some exceedingly well-trained Malamutes, Akitas, and Rottweilers. Thus, it wasn't their breed that led these dogs to have low rankings, but ineffective training techniques.

Nonetheless, it's well worth considering *your* training skills and how well they meet the needs of your dog. Dogs bred to hunt, herd, work, and compete are high-drive dogs. If you're going to buy a Ferrari, you had better know or learn how to drive one.

Behavior Is Changed by Its Consequences

In the wild, when young animals investigate the real world, the consequences of their actions range from exceedingly pleasant to downright unpleasant. Nature can be a very rewarding or a very strict teacher.

Animals learn from their actions and modify their future behavior accordingly. Most young animals embrace their physical and social environment with gay abandon. Many behaviors have pleasant and rewarding outcomes, and so these increase in frequency, whereas behaviors that have unpleasant consequences are strongly inhibited, such as ploughing through rosebushes or eating a wasp.

As animals learn, safe-and-rewarding, environmentally appropriate behaviors are repeated, whereas dangerous, unpleasant, environmentally inappropriate behaviors quickly decrease in frequency. As a result, "punishment from the environment" becomes less and less frequent.

However, even though "punishment from the environment" can sometimes be harsh, there's no reason for us to follow suit. Both the environment and computers in laboratories have the same enormous constraint when teaching/training animals; they cannot speak to the animals to inform them which red berries to eat and which red berries to avoid. We can.

In 1898, Edward Lee Thorndike suggested a "law of effect," which states that "responses that produce a satisfying effect in a particular situation become more likely to occur again *in that situation*, and responses that produce a discomforting effect become less likely to occur again *in that situation*." Basically, what this means is, behavior is changed by its consequences. Also notice that Thorndike had cottoned on to the fact that the effectiveness of learning is specific to the situation. This is true in nature and when we teach our dogs.

The terminology has changed over the years, but the mantra remains the same. Responses are strengthened and reinforced by pleasant consequences, and they are weakened and inhibited by unpleasant consequences. When it comes to dog training — or any educational or motivational scheme involving children, students, employees, colleagues, or clients — this has led to the familiar binary feedback of *rewards versus punishments*.

Defining Rewards and Punishments in Dog Training

In everyday life, we define rewards and punishments by their intrinsic nature: A reward is something that's pleasant, and a punishment is something that's unpleasant. However, in science as well as in dog training, these terms are defined, not by their nature, but by their *effect on behavior*. Thus:

A *reward* is defined as *any stimulus* that causes the immediately preceding behavior to *increase* in frequency, and thus be *more likely* to occur in the future in that situation.

A *punishment* is defined as *any stimulus* that causes the immediately preceding behavior to *decrease* in frequency, and thus be *less likely* to occur in the future in that situation.

This can be a tricky notion to grasp and can lead to many misunderstandings about what makes an effective "reward" and what makes an effective "punishment." Yes, food rewards are pleasant, but the nitty-gritty is *how well do they reinforce behavior?* Similarly, leash corrections are unpleasant, but *how well do they inhibit behavior?*

Do All Pleasant Stimuli Reinforce Behavior?

Usually yes, but not nearly as much as they should. Although pretty much everything we regard as a typical "reward" for a dog is normally intrinsically pleasant, such as praising, petting, and offering food, few people praise convincingly or sufficiently, the effectiveness of petting depends very much on the dog's perception of the person, and food rewards are often used in ways that substantially decrease their effectiveness. Most people offer *far too many* of the *same-old* food rewards and deliver them on a *consistent and predictable basis*, so they become taken for granted and lose their reinforcing power.

In upcoming chapters, you'll learn how to motivate dogs using considerably more powerful reinforcement schedules; power up praise and food by increasing their *secondary* reinforcing potential for classical conditioning; and largely replace food rewards with much more powerful life rewards to reinforce desirable behavior and manners. Moreover, an integral aspect of lure-reward training is regularly testing response reliability to objectively evaluate the effectiveness of training and offer *proof of progress*.

Do All Unpleasant Stimuli Inhibit Behavior?

They should, but when applied by people, they usually don't. In chapter 14, you'll learn that for any aversive stimulus to effectively inhibit undesirable behavior, its application must adhere to six stringent criteria. For example, punishment must be consistent, immediate, and instructive. Most people are rarely consistent or attentive 24/7, so it is next to impossible for people to punish effectively.

An unpleasant stimulus only qualifies as a punishment if it decreases the future incidence of the targeted behavior. If the stimulus does not inhibit behavior, then it is *not* functioning as a punishment. For example, if after multiple leash corrections, a dog still pulls on-leash, then leash corrections cannot reasonably be defined as punishment because they are not working. They are *not* corrections. The leash corrections are certainly unpleasant, but that's all they are — unpleasant stimuli.

At best, consistent and well-timed aversive punishment only communicates, "Stop what you're doing," and at the worst, "No," "Stop everything," or "Freeze." A huge reason that aversive stimuli are largely ineffective is that they lack instruction and do not inform animals *which* specific behavior to stop or *what* to do instead. Also, without an instructive warning prior to punishment, the animal has no means to avoid the aversive stimulus. *Comprehended verbal warnings* are considerably more instructive than beeps and buzzes.

Of course, all this begs the question: If we are going to offer verbal instruction beforehand, need punishment even be aversive, or is punishment even necessary?

Need Punishment Even Be Aversive?

If something neutral, or even pleasant, decreases the frequency of undesirable behavior — for instance, instructing a dog to Sit when they are jumping on a visitor — then the request to Sit is functioning like a punishment by inhibiting jumping up. In fact, Sit is the very best "*nonaversive* punishment" in that situation, as it is in many others.

The gleaming gems throughout this book are the many nonaversive techniques for preventing and resolving misbehavior and noncompliance without ruining a dog's day or yours. To me, aversive techniques are mostly ineffective, never sufficient, often counterproductive, and therefore unnecessary and unwarranted, especially given the wide range of much more effective nonaversive techniques at our disposal.

Why scare a child who's trying to learn calculus? Calculus is sufficiently scary by itself. Yes, the child will make mistakes, but they are trying their best. And while learning, dogs will make mistakes, too.

But why treat our best friend like our worst enemy under the guise of training? For example, if we can get our dog to stop barking by teaching them to Shush on cue, why do anything else?

I am a fervent believer that training should be quick, easy, effective, and of course, enjoyable for both dogs and people. To me, there is simply no reason for training to ever be discomforting, unpleasant, scary, or painful.

The very heart of lure-reward training comprises (1) verbally cuing and reinforcing desired behaviors by providing rewarding consequences — oodles of praise and life rewards, such as interactive games, walks, and dog-dog play; (2) testing response reliability to confirm comprehension and motivation; and (3) preventing and eliminating undesired behavior via *instruction and guidance*, that is, *using words.*

The Super-Secret Sauce: Increasing the Good Decreases the Bad

Every great dish has a secret sauce, and here is the one for lure-reward training: Reinforcing desirable behaviors (using rewards) creates a positive feedback loop that gets stronger and stronger over time. As desirable behaviors increase in frequency, they are more likely to be reinforced, and this further increases their frequency. Since there are only so many hours in each day, undesirable behaviors naturally get crowded out.

For example, rewarding a dog for calmly lying down by your side in your home office *increases* the time that your dog will lie down by your side, which has become the current treat zone. This obviously *decreases* the time available for the dog to practice agility in the living room, bark at passersby at the front window, or chew the remains of the remote control on the remains of the couch, that is, the couch with all the urine stains.

Here's another wonderful example: Every time you reward a dog with three pieces of kibble for peeing *on cue* in their toilet area, that increases the likelihood the dog will pee in their toilet area when requested — so they may cash in their urine for kibble. And once your dog routinely pees in the toilet area, there is less urine for the couch. A dog only has a finite daily supply to go around, which is why rewarding a dog for emptying out in the toilet area is the best (quickest) way to prevent, *and resolve*, housesoiling.

Moreover, encouraging good behavior is a darn sight easier on the brain than developing the exquisite timing and absolute consistency that are required to discourage undesired behavior with punishment.

Here is one of the many delightful aspects of lure-reward training: Lousy timing seldom damages effectiveness. After just a few rewards, dogs will be closer, calmer, and quieter, for longer and longer periods. Moreover, we may be utterly *inconsistent* when dishing out rewards. Inconsistency and unpredictability increase the dog's anticipation, which increases the reinforcing power of each reward.

Praise Your Dog for Being Good!

In a dog training class, I always tell people, "Praise your dog! Reward your dog!" But people often reply, "But why? She's not doing anything." Ahem! Precisely! That's the *whole point*! Your dog is not doing anything that irks you, like running around, barking, jumping up, climbing the walls, trotting along the kitchen counters, and so on. Dogs do so many things that engender wrath in their owners, so why don't we praise and reward them for *not* misbehaving?

Try this quick experiment. Put a baggie of kibble in your pocket, sit down, and watch your dog. Every fifteen seconds, classify your dog's behavior as either "desirable" or "undesirable" and keep score on a piece of paper. Remember, "desirable" also includes your dog not doing anything that annoys you.

Ignore misbehavior but praise your dog and offer a piece of kibble every time you classify their behavior as desirable. Either walk over to the dog to offer the kibble or let the dog come to you. After just ten minutes, you'll notice a few things: Desirable behavior routs undesirable behavior by a huge margin; your dog is most likely sitting or lying close by and keeping an eye on your every movement; and the time spent sitting, lying down, and watching you have all increased dramatically.

Dogs — as well as children, spouses, and employees — are good most of the time and yet ignored. However, no matter how good, your dog will become even better when you focus on their behavior and offer the *occasional* reward for not misbehaving.

Focusing on the good is the most effective way to change, improve, and fine-tune behavior and temperament — of any animal, people included, *especially yourself.* By so doing, good behavior increases in frequency and undesirable behavior is squeezed out, and punishment becomes simply... unnecessary. Rewarding your dog for desirable behavior should always be the prime directive.

Dogs Attend Preferentially to Body Language, Tone, and Volume

Cartoonist Gary Larson was so absolutely spot-on with his animal cartoons. My favorite is the classic panel of a dog being scolded by her owner and hearing only, "Blah! Blah! Blah! Ginger! Blah! Blah! Blah!" Most of the words we use are irrelevant to a dog and are consequently ignored. Just very occasionally a word matters to them and is worthy of their immediate response, such as their name and words like *dinner* or *walkies*. During training, the prime directive is turning just a few human words, our instructions, into the *most* salient aspects in a dog's life. Some examples are "Come," "Sit," "Down," "Shush," "Go pee," and so on. Other than those commands, our endless chatter will likely remain, in the words of Perry Mason, "incompetent, irrelevant, and immaterial."

So many stimuli are bombarding a dog's brain at any given moment, and naturally, dogs attend only to the most dog-relevant stimuli, such as smells, body contact, movement, other dogs, and so on. Human logorrhea is not dog-relevant, which makes it a challenge to teach a dog that there are a few human words that are in fact highly urgent and important for them. How do we do that? Teaching dogs ESL is the quintessence of lure-reward training.

People normally pay attention to the spoken word, and unless they are interrogators or poker players, they miss most facial and body language "tells." Dogs, on the other hand, always pay more attention to facial expressions and body language than to *what* we say. As we're trying to teach the *meaning* of verbal requests, the dog is often preferentially responding to our tone and volume, or especially, minor movements. Many dogs teach themselves that if the owner is standing, facing them, and looking at them, it's a good plan to sit.

At a glance, a dog can tell more about us and our mood than perhaps we ourselves, our mother, or our psychologist. Even when dogs don't understand our actual words, they are sensitive to the nature of our voice. Many people tend to modulate their voice depending on the instruction, generally using a soft, sweet, lulling voice to seduce their dog to come and follow, a sweet voice with rising intonation when asking a dog to sit, and a louder, gruffer, more authoritative voice to get their dog to lie down and/or stay. Similarly, people tend to raise their head when saying, "Sit," and they lean over when requesting Down.

Although changing how we present an instruction might *temporarily* change behavior in the moment, it is unlikely to cause a permanent change in compliance or reliability. Yes, our dog *might* comply right now if we sweetly implore or gruffly shout, but in the future, our dog might disregard requests, especially if given in a different tone. For example, if we laugh while instructing Down, our dog might just laugh back, since they are relying on tone and posture for interpretation.

I always recommend that people stand still and use a neutral tone when teaching the meaning of instructions, so the dog may focus on *what* we say, and so the dog's compliance does not become contingent and dependent on tone, volume, any postural change, or hands-on contact.

The nitty-gritty of communication is teaching dogs the meaning of *words*, so that dogs understand our requests even when we are out of sight, in a different room, at a distance, and especially, if the dog's back is turned. The ideal is for dogs to come when we say, "Come," even when we can't see our dog and the dog can't see us.

Dogs Thrive on Clear Communication

Training dogs is not that complicated. I've always maintained that dog training is not rocket science, but it *is* science. Well, science and soul combined.

Even so, we can't just tell a dog what to do without first teaching ESL — English (or Spanish, Japanese, or Hindi) as a second language. Training is all about clear communication.

Among themselves, dog-dog communication is very clear. Even humans can read the messages: *I wanna PLAY! Don't do that! Quit it!* In fact, dogs "speak" using many languages — auditory, olfactory, and especially, body language. However, just as dogs can't mimic human speech, we would fail miserably if we tried to communicate with a dog by woofing, growling, moving our ears, or by bending down and waggling our tailless backsides to solicit play.

The first goal of any training technique is to teach dogs the meaning of the words we use for instructions, requests, and extra guidance should they go off track. Remember, you're not teaching your dog to sit; your dog knows how to sit. You are teaching your dog to Sit *on cue*, so that you may communicate *when* to sit, *where* to sit, and *how quickly* and for *how long* to sit.

Opening communication channels by teaching dogs ESL is transformational for dog training. Teaching a dog to respond on cue enables you to teach your dog more easily (and quickly): *when* and *where* to pee, *what* to chew, *when* to shush, and *when* and *where* to settle down. And teaching dogs to jazz up and settle down on cue enables you to let your dog know *when* and *where* it's cool for them to let off steam.

Clear verbal instruction is essential for cuing basic manners, but where verbal instruction excels is providing additional guidance when dogs err. Precise instructions such as Steady (to slow down), Hustle (to speed up), Shush, and *Tranquilo* are far more effective than shouting a blanket "No!" or giving physical corrections. *Verbal* guidance communicates two vital pieces of information: (1) Stop what you're doing, and (2) *do this instead.*

Once dogs begin to understand more and more of our vocabulary, it is possible to communicate via grammatically correct sentences. Here are my two favorites: "*Claude*, *find* your *Kong*, *go to* your *bed*, *settle down*, and *shush*." (Italicized words are the ones Claude understood.) And: "*Phoenix*, *come* here, *take* this, and *go to . . . Jamie*. Please." Whenever I said this, Phoenix would come and sit (I define Come as Come-Sit-Stay-Watch), and I would give her a note for Jamie, then off she would run, hunting high (in his room) and low (in the garden), and then sit for delivery. We called this routine Malamute Mail.

Define What You Mean by "Sit"

Before teaching dogs the meaning of our instructions, we first need to precisely define for ourselves what we mean by each word. Take, for example, the instruction Sit. How can a dog get that wrong? Amazingly, though, very few *people* have a clear picture in their head regarding what they would like a dog to do when asked to sit. So dogs improvise, which owners view as "making mistakes," and what should have been an extremely quick-and-easy training exercise becomes a time-consuming annoyance.

In seminars, I often ask audiences to define what they mean by the instruction Sit. Most people shout out, "Put your butt on the ground!" So I ask everyone to write their definition on a piece of paper. Then I choose a volunteer and tell them, "I am your dog. I am perfectly trained and 100-percent reliable. I will do exactly what you ask based on your definition." Then I ask the person to give me a few seconds to wander around the room and get distracted, and then say, *just once*, "Ian, Sit," and I will follow their instructions to the letter.

Few people follow my instructions. One lady said, "Ian, Sit." Then when I didn't sit immediately, she said, "Sit! Sit!!! SIT!!! Ian put your butt down," and then she ran up to me, said, "Baaaaad Ian," and swatted me on the butt.

"Hey!" I said, "What are you doing? I thought I asked you to only say, 'Ian, Sit,' *once*."

"But you're *not* sitting."

"I know. I was just going to, but after I finished chatting to this lady in the front row." I waved the definition in front of their face. "Look, you only asked me to 'put my butt on the ground.' You didn't tell me *when*. I had no idea the sky was falling. Would you like to amend your definition?"

The woman revised her definition to read "Put your rear on the ground NOW!" I told her we'd try again, and after a couple of seconds of me wandering, she said, "Ian, Sit." This time I sat like greased lightning, but then I immediately jumped up and ran over to a man at the back of the room and started faux humping. (The man had agreed to be humped, by the way; I had arranged his consent prior to the lesson.) I looked back over my shoulder, smiling, and said, "I presume you would like to amend your definition once more."

This time, she changed her definition to "Immediately, put your butt on the ground, and keep it there until the next instruction."

This demonstrates that the definition of even the simplest instruction usually conveys at least three pieces of information. For example, with Sit, we define the *body position* (butt on the ground), *when* to adopt this position (immediately), and *how long* to hold the position (until the next instruction). Defined this way, Sit makes the instruction Stay unnecessary, since our definition of Sit means "immediately Sit-Stay."

When teaching puppy classes back in the 1980s, I gave everyone a leather-bound notebook and asked them to write down every word they planned to use to communicate to their dog and then to define each one — a Doggy Dictionary. When owners got frustrated in class, I would often check their Doggy Dictionary and explain that the dog was doing exactly what they asked. For example, we were once practicing commands with owners sitting in chairs. One owner instructed her dog, "Down," and then got upset when her little dog jumped into her lap and lay down. But her definition did not specify lie down *on the floor*.

Lack of Comprehension Is the Major Reason for Noncompliance

When dogs are noncompliant, rather than reevaluating their training, people tend to blame the dog and accuse them of being stubborn, stupid, dominant, or even intentionally dissing them. I doubt this is ever true. The number-one reason for noncompliance is a lack of comprehension.

Whenever I say this, I'm always met with disbelief: "He knows what it means!" However, one of the secrets to lure-reward training is that *the dog's comprehension is tested* every step of the way, and as you'll discover, these absolute scores are very revealing and, indeed, can be quite a shock. We continually evaluate our dog's progress so we can confirm where training is succeeding and where it is failing to communicate exactly what we want. Then we adjust training to fix the cracks in our communication.

With dogs, as with people, comprehension is not either/or. It is so seldom that the dog "knows it" or "doesn't know it." Rather, their understanding is somewhere in between — incomplete, ever-changing, and hopefully, progressively improving. A common reason is, like the Sit example above, we haven't clearly and specifically defined and communicated what we mean by a request. A more common reason is that a dog's reliability varies considerably according to the trainer, the location, the distance, and whatever else is going on. A dog taught Sit by one person in the kitchen won't instantly understand that "Sit means sit" wherever they are, whoever says it, and especially, when some distance away.

I use a very simple index — a *response-reliability percentage* — as a combined measure of both comprehension and motivation. The score represents a dog's reliability when responding to handsignals or verbal instructions. Dogs seldom score zero percent because they teach themselves by "reading" us. And dogs only score 100 percent over short-term testing. No dog or human is perfect. We might want our dog to be perfect, but that's as elusive as perfecting a golf swing or a soufflé. Imperfection is what makes golf, cooking, and dog training so wonderful.

The scores, though, can be a brutal wake-up call for many owners and obedience competitors. I can remember testing an Obedience Trial Champion sheltie in Denver. In the owner's kitchen, the dog scored 96 percent; in the living room, 82 percent; in the backyard,

48 percent; and in a dog park, 17 percent. Yet in the trial ring, the dog had two back-to-back perfect 200 scores (100 percent).

Of course, the sheltie was a professional competitor and would relax at home but turn it on when it was showtime. That is the lesson for us lesser mortals: We must have a thoroughly realistic view of our control over our dog before walking from house to car off-leash or letting the dog off-leash in unfenced areas. Teaching our dog to generalize commands and have high reliability when we need it, while also allowing our dog to chill and relax most of the time at home, is easy to do and involves simple exercises you'll love.

One thing this hopefully makes clear is that no dog learns everything in a single session. Rather, we spend our life with our dog progressively solidifying and testing comprehension, while also motivating our dog to *want to learn* and do what we ask. The lure-reward motivation program is beyond any other. The goal of training isn't only teaching the meaning of words; it's to make training so enjoyable, so much fun, that dogs want to play the training game with us. Training simply must integrate a dog's favorite activities, such as walking, sniffing, dog-dog play, and interactive games with people, until training and play become indistinguishable. When that happens, engagement and responsiveness skyrocket.

Eventually, our dog will promptly, willingly, and happily comply because they want to, whereupon external rewards are no longer necessary because our dog has become internally reinforced and self-motivated.

The Importance of Follow-Up

As puppies become adolescents, they tend to ignore more and more of our requests and commands. One obvious reason is that they develop other doggy interests and activities, which compete with us for our dog's attention and time. However, there is an additional reason. Most of the time, after issuing instructions, *owners don't follow up*, and so the dog learns that the "commands" have become irrelevant. Basically, many owners unintentionally train their dogs not to respond, and the dog's lack of response effectively crumbles their person's confidence as a trainer, and so they become even less inclined to follow up.

When, on occasion, owners do follow up, it is often with impatience, frustration, or anger. And so dogs learn that their person is inconsistent, unpredictable, and prone to temper tantrums, which puts them on edge.

Dogs, being fast learners, try their best to read us, and they can usually tell whether we will follow up or not. For example, if we are sitting down, leaning back on the couch, with a drink in one hand and our phone in the other, and the television is on, our dog can calculate a 99.99 percent probability that nothing we say will be relevant to them for the rest of the evening. Nor, if we do ask the dog to do something, will we follow up and make sure they comply. Whereas, when we are standing and staring at the dog when giving a command, we probably will insist that the dog comply.

This is the *consistency conundrum* that bedevils all training. However, this is where lure-reward training literally shines with its wealth of words. In part 4, I offer a variety of simple, dog-friendly techniques to resolve misbehavior, increase reliability, and create extremely high levels of compliance on demand. But first, it's time to teach your dog a few words.

Part 2

LURE-REWARD TRAINING STAGE ONE: INSTRUCTION

Lure-reward training is a comprehensive training program with three stages:

- **Instruction:** Using food lures to teach handsignals, and then using handsignals (without food) to teach the meaning of verbal requests so dogs understand *what* we would like them to do.
- **Motivation:** Initially using food rewards to teach dogs to *want* to do what we would like them to do, and then primarily using much more powerful *real-life* rewards.
- **Compliance:** Using our *voice* for instructions, for guidance to get dogs back on track when they misbehave, and for increasing compliance with our verbal requests.

Part 2 is focused on stage one: instruction. It explains how to teach dogs ESL to open up communication channels. First, we use food lures to teach dogs the meaning of handsignals, or more specifically, *which* body position or movement is required following each handsignal. Once dogs understand handsignals (without food in the hand), we use the handsignals to teach the meaning of each verbal instruction, that is, which body position or movement is required after each verbal instruction.

As we steadily expand our dog's vocabulary, we routinely test the dog's comprehension, which offers proof that training is both successful and progressing. Ultimately, we can indeed talk to our dogs *and* our dogs can understand us.

Chapter 5

An Overview of Lure-Reward Training

My working definition of a trained dog is one that is under reliable, off-leash, verbal control, including when at a distance, when distracted, and without the continued need of any training tool whatsoever.

Dog training changes the frequency, timing, and quality of behaviors. It increases the frequency of some behaviors, decreases the frequency of others, and changes their situational appropriateness by putting selected behaviors on cue.

- **Increasing the frequency of desirable behaviors:** This includes approaching, staying close, following, being quiet, remaining calm, and chewing chewtoys.
- **Decreasing the frequency of undesirable behaviors:** This includes housesoiling, destructive chewing, barking, jumping up, pulling on-leash, and hyperactivity.
- **Putting desirable behaviors on cue:** This can include almost any request, such as Come, Sit, Down, Stand, Stay, Rollover, Heel, Hustle, Fetch, Find, Go pee, Go to bed, and so on.
- **Putting the cessation of undesirable behaviors on cue:** This includes requests such as Settle, Shush, Steady, and Off.
- **Improving the quality of desirable behaviors:** This means straighter sits, faster recalls, and stylish heeling. Why? Why

not? It's just as easy to train a dog to ooze panache as it is to allow a dog to be sloppy.

I generally qualify dog behavior as either "desirable" or "undesirable." By that I mean, whether it is acceptable and appropriate or unacceptable and inappropriate to the dog's human companions. Some people prefer to describe behavior as "good" or "bad." I'm OK with "good" because I truly believe that when we acknowledge a dog's needs, feelings, and point of view, and do so with sufficient motivation, many dogs do develop a "desire to please." But "bad" assumes that dogs are intentionally acting out or misbehaving to get our goat and annoy us. Whereas this may be true of some people, I don't think this is true of dogs. (Well, except for a few breeds that have an outrageous sense of humor.) From most dogs' perspective, they probably view what we'd call "good" behavior as good, but "bad" behavior as even better. That is, what we don't like and might consider a "behavior problem" is usually what dogs really enjoy doing.

Lure-Reward Training: So Old It's New Again

Lure-reward training was specifically designed primarily for teaching animals to respond reliably on verbal cue. Luring using food is not a new training technique. It's been around forever. I was fortunate to grow up watching my dad and grampa lure their gundogs, my grandmother lure our fourteen farm cats, and the farmhands use food lures to teach very basic verbal instructions to farm animals, such as "Com'ere" and "Back up" and destinations like "Stall" and "Pasture."

As a kid I was amazed that the animals could understand English. I was always mightily impressed when I would see a herd of cows walk from the pasture to the milking parlor, each one going into its appropriate stall. Of course, learning which stall is yours is easy from a developmental perspective; when a young heifer joins an established herd, there are only one or two empty stalls to choose from, and only one will have hay or silage to munch on.

Verbal control fascinated me. When riding my favorite heifer,

Blackie (a silly name, really, because half the herd was black), we taught her "whoa," "go on," "right," "left," and "graze." I would give instruction from atop her withers, and my younger cousin, Paul, would follow the instructions with a cabbage in his hand, periodically peeling off a leaf at a time to lure Blackie to follow him. I used to love pretending that I was steering Blackie by her ears: both forward for go, both back for whoa, left forward and right back to turn right, and right forward and left back to turn left. Of course, I knew even then that she was just following the cabbage, but it was wonderful to dream.

And then dreams came true. Blackie began to anticipate. For example, she would move right when I said, "Right," and moved her ears accordingly — because she had learned that the verbal instructions and ear signals predicted that the cabbage was about to move right (because Paul was following my instructions). This was lure-reward training, pure and simple. And to this day, I still think it's magic!

Later in life, when I taught the ten-week course on dog behavior at the UC Berkeley Extension, I included one week on dog training. In preparation, I read two books by Dr. Leon Whitney — *The Natural Method of Dog Training* and *Dog Psychology: The Basis of Dog Training.* I loved these books because they talked in detail about using food lures and rewards for off-leash training... for beagles! And he had a sister book for cats. Everything was so familiar; I just gave the technique a name for the lecture. People were amazed. I simply had no idea that these techniques were so different from the default on-leash training techniques of the day.

Lure-reward training teaches dogs *what* we would like them to do by using food lures to teach the meaning of handsignals and, eventually, verbal requests. Once the dog learns the meaning of handsignals, food lures are dispensed with, and handsignals (without food) are used to solidify comprehension of verbal requests. Verbal instructions are by far the most practical way of giving instructions.

Further, while food rewards are initially necessary to motivate dogs to *want to do* what we would like them to do, food rewards are gradually phased out and replaced by much more powerful life rewards. Eventually, the method eliminates the need for any external reward, since dogs become internally reinforced and self-motivated — a dog *wants* to play/train with us because "just doing it" is more than sufficient reward.

Quick, Easy, Effective, and Enjoyable

Lure-reward training is my method of choice because it's easy to learn, produces rapid results, is effective at modifying behavior, and is enjoyable for dogs and their people. Learning this method is as easy as 1-2-3-4:

1. Request → 2. Lure → 3. Response → 4. Reward

The lure functions like an olfactory halter; as you move the lure, the dog's nose follows, as does the required behavior. Further, phasing out food lures and the reliance on food rewards is considerably easier than trying to phase out food with other training techniques, such as clicker training, or to phase out reliance on physical training tools, such as leashes, halters, harnesses, training collars, and physical prompts. Additionally, lure-reward training *off-leash* is a quicker one-step process, whereas training a dog on-leash is a two-step process: first on-leash, and then off-leash.

Lure-reward training is truly magical in terms of the speed of results — the quickest training technique on the planet for teaching reliable, off-leash, cued responses, especially from a distance. The essence of all reward-training techniques is *reinforcing desirable behavior*, and the quickest way to do that is to lure the behaviors that you want to reinforce, so that you can reinforce them as much as you like. In fact, you can lure multiple behaviors in rapid succession and reinforce at will.

The Three Stages of Lure-Reward Training

Lure-reward training is the only reward-based technique that maximizes the initial use of temporary training aids (lures and rewards) and then formalizes the process of phasing them out, so that they don't become permanent management tools (crutches). Also, lure-reward training regularly quantifies reliability and compliance — off-leash and at a distance.

Training unfolds over three distinct stages: (1) instruction, (2) motivation, and (3) compliance. Roughly speaking, stages one and three each take up about 5 percent of training time, while motivating dogs to respond reliably, willingly, and promptly comprises a full 90 percent of the process. The three stages are explored fully in parts 2, 3, and 4.

Instruction

The first stage of lure-reward training is teaching dogs *what* we would like them to do. Actually, more specifically, it is teaching dogs the *meaning of the words* we use for instruction. It's no use shouting if a dog doesn't understand the meaning of verbal "requests." Moreover, dogs can never be reliable off-leash without verbal comprehension. Teaching dogs the meaning of human words is the essence of successful dog training.

A key element is confirming comprehension *by phasing out food lures*. First, we practice luring until we are proficient, and the dog is focused on our hand movements, then we lure with an empty hand. Voilà! We've taught the dog handsignals! Easy-peasy. Then, we use handsignals (*with no food in our hand*) to teach dogs the meaning of verbal requests. We can identify when a dog understands the meaning of a word by watching for the first time that the dog responds *after* the verbal request but *before* the handsignal. When this happens, congratulations! Your dog is beginning to learn the meaning of your words.

It is important to have a go at *phasing out food lures during the first few sessions*. Otherwise, food lures will become *bribes*, and nothing undermines training faster. I cannot emphasize this strongly enough. In a nutshell: *Food lures* are used to teach an *attentive and willing* dog *which specific behavior* is desired, whereas *bribes* are used to coerce an *unfocused or unwilling* dog to pay attention by promising the food in our hand as a reward.

As soon as possible, get the food out of your hand and into your pocket or bait bag, or on a table, to be used only as rewards after reward-worthy responses. It is so important to prevent a dog's responses from becoming dependent on the trainer having food in their hand. The earlier you try this, the easier it is. Ultimately, once dogs learn the meaning of handsignals and verbal requests, food lures are unnecessary and should be dispensed with entirely.

A vital part of instruction, which applies to all stages of training, is to regularly check your dog's comprehension of handsignals and verbal cues by testing and tracking response-reliability percentages to evaluate the speed of progress (see chapter 7). As soon as you begin to quantify your efforts, your skills as a trainer, your motivation, your dog's motivation, and your dog's reliability will all take a delightful zoom forward.

Motivation

The second stage of lure-reward training comprises motivating dogs to *want* to do what we would like them to do. Food rewards are an ideal motivational tool, as they are intrinsically savored by most puppies and dogs, and so they are especially useful at the start of training and with owners who have difficulty praising convincingly. Using kibble also makes it easy to vary the quantity and quality of rewards to match the speed and quality of responses. Food rewards quickly teach dogs to love training *and the trainer*, which is especially useful when unfamiliar people work with your dog. Using food rewards gives everyone a racing start.

However, in the very first training session, start reducing the number of rewards you use by doing the following:

- **Ask more for less:** Progressively ask for more and more behaviors and longer and longer duration behaviors for fewer and fewer rewards. Keep track of the response:reward ratios (see also chapter 7) as you do this. It will amaze you what a puppy will do for the prospect of a single food reward.
- **Make reinforcement more effective:** Once you have reduced

the number of food rewards, you can make them appear to be random, and thus be *unpredictable*, which significantly increases their reinforcing value, and you preferentially reward the dog only for better-quality responses.

- **Replace food rewards with life rewards:** Dogs are much better motivated by the activities they love, such as walking, sniffing, playing with dogs, and playing interactive games with you, such as fetch, tug, tag, and hide-and-seek.
- **Put "problem" behaviors on cue:** People consider some dog behaviors as "problems," such as barking, jumping up, and hyperactivity. However, when you put these behaviors on cue, they, too, may be used as extremely powerful life rewards because they represent what dogs really want to do.

Ultimately, dogs will become *self-motivated*. Dog-dog play and interactive games are perhaps the best life rewards in training. By integrating very short training interludes within lengthy play sessions, we convert a potentially catastrophic distraction to training (playing with other dogs) into the very best life reward to reinforce training — "Go play." In fact, when dogs are playing with other dogs, pretty much the only effective reward at our disposal is letting our dog *resume play* with other dogs. Succeed at turning dog-dog play into a reward and you are on the fast-track to success.

Even though, eventually, we very seldom use food rewards to reinforce basic manners, we may give our dog an occasional food reward anytime we like, and of course, we must continue to offer food rewards willy-nilly for classical conditioning, such as when encountering another dog or person on the sidewalk.

Compliance

The third stage of lure-reward training is building reliability and increasing compliance. The sheer beauty of lure-reward training is that, with our dog's enhanced vocabulary, we may offer additional guidance when dogs err.

Often, that can be done with a single word, like Sit, which can effectively communicate two or three highly specific pieces of information in an instant: (1) Cease what you are doing, (2) do this instead, and (3) when required, the degree of danger of noncompliance.

For example, if a dog is barking for longer than is acceptable to us, we can simply instruct Shush. If our dog is about to pee indoors, instruct Outside or Toilet. If our dog is chewing our slippers or underwear, instruct Chewtoy. If our dog is counter-surfing, instruct Off. If our dog is running around the house, instruct Bed and Down. And if our dog is doing most anything else that bugs us, such as humping another dog, chasing the cat, about to run out the front door, or jumping up, instruct Sit.

Achieving reliable compliance when we need it involves giving our dog at least two names: one is the dog's nickname, which is used when offering "suggestions." That is, these are instructions we'd like the dog to follow, but if they don't, it's no big deal. Then we teach our dog a different "formal name," which we use to signal a "formal command" — one that must be followed immediately with no discussion. This formal name is taught by practicing repetitive reinstruction until compliance. Without using force, we calmly and persistently insist that our dog comply before we say, "Free dog," "Go play," "Go sniff," "Fetch," "Tug," or "Let's go."

Ultimately, we want to train a dog to comply even in the most distracting scenarios, such as when they are off-leash at a distance and in an emergency when safety (of the dog and/or people) requires instant obedience. Distance commands, especially emergency Sits or Downs are the foundations of off-leash reliability that give dogs considerably more supervised freedom.

Defining Lures, Rewards, and Bribes

The only Achilles' heel of lure-reward training happens if people do not phase out food lures in the first few sessions. As soon as dogs approach adolescence and become distracted by other doggy interests, food lures often become bribes to try to get them to pay attention and mind. And as parents and politicians quickly learn, bribes are not entirely effective.

Initially, food is used as both lures and rewards, and these are easy to distinguish. However, bribes are an elusive shape-shifter that are more difficult to define.

Food Rewards Reinforce Behavior and Associations

Rewards are given *after* a behavior in order to *reinforce the behavior* so that it is *more likely to occur in the future* in that situation. In other words, a food reward is meant to be a pleasant consequence when a dog does what we ask, but it's only given *after* the dog does what we ask. In lure-reward training, rewards also *strengthen the associations* between the verbal request, the following handsignal, and the desired response. Moreover, dogs quickly associate food rewards with training and the trainer.

Food Lures Teach Willing Dogs What to Do

Food lures are presented *after* a verbal request and handsignal, but *before* the behavior, to teach an *attentive and willing* dog *which specific behavior is required.* Initially, food in the luring hand focuses the dog's attention on the hand movement. Then, once the dog understands that the hand movement is actually a specific handsignal requesting a particular behavior (like Sit), then the handsignal can be used and will work with no food in the hand.

A lure is a teaching tool that's used to foster communication. Unlike a reward, it is not intended to reinforce the behavior. In fact, the food in the luring hand is never given as a reward. Food rewards are given only from the nonluring hand. In essence, the food lure is only needed to teach the dog the meaning of each handsignal. The premise is that an attentive and willing dog will gladly do anything to get the food if only they knew *what* we want. Once dogs know what we want (either using a handsignal or verbal request), the food lure is no longer necessary.

Bribes Are Used to Coerce Unwilling Dogs

A food lure becomes a bribe when it is used to try to *coerce*, convince, or motivate an *inattentive or unwilling* dog to pay attention and mind. That is, a bribe is presented *before* the behavior (like a lure), but now it is the "promise" of the food reward the dog will get *afterward* if they do what we want. However, if a dog hasn't yet learned handsignals, it will still be a mystery what exactly we want them to do.

A bribe is neither an effective lure nor an effective reward: A bribe doesn't teach the dog which specific behavior is required, or the meaning of a handsignal. Instead, it's more like a desperate request to get a dog to please, pretty please, pay attention and acknowledge our existence. Similarly, a bribe is not an effective reward, since it is presented *before* the behavior. Behavior is *only* reinforced (or inhibited) by the consequences that occur *after* the behavior.

Suffice it to say, it's critical to phase out food lures as soon as possible, so they don't become bribes, and in chapter 6, I suggest several ways to accomplish that.

Food Lures Have a Half-Life

A lot of people think that the magical effect of a food lure is that they can get their puppy to gleefully pop up and down between Sits and Downs, over and over, *ad infinitum*. And yes, food lure–driven, whack-a-mole displays are impressive and delightful to watch. But trust me, this is temporary. *Food lures will only capture a dog's attention for the first hundred or so repetitions*. Luckily, most dogs only need six to twenty repetitions to learn what a handsignal means, since they are fluent in body language.

Food is a very convenient and effective tool at first, but the attraction of a single piece of food in your hand quickly wanes, especially if the dog is getting a hundred pieces of kibble every day for free in a bowl at 5 p.m., plus maybe tastier treats elsewhere. Similarly, as a puppy grows older and discovers other doggy interests, those interests will become far more interesting than a piece of kibble or even a tasty treat in your hand.

Remember, food lures are not intended to motivate a dog prior to the response (in which case, the lure is being used as a bribe). Lures are used solely to teach which response is being requested. The food focuses the dog's attention on the hand movement so they learn the meaning of handsignals. Once a dog understands handsignals, my advice is to go cold turkey on food lures as quickly as possible before they exhaust their half-life and their magic.

Which Type of Food?

By and large, food lures and food rewards work brilliantly in dog training, especially during the initial stages. Using food as a lure to teach handsignals and ESL is the best way to get young pups or older dogs (that are new to training) up and running as quickly as possible. Meanwhile, food is a great initial reward for most dogs, and kibble is the most convenient and easiest to use, especially when children or strangers work with your dog.

The Best Food for Training

Dry kibble is just so convenient to use as lures and rewards, whereas canned food and raw diets are much less so. Certainly, in the past, there was an enormous selection of not-so-healthy commercial kibble in the marketplace. Now, though, there is an impressive selection of high-quality kibbled food in terms of quality and balance of ingredients. My preference has always been for air-dried foods with no preservatives and no denaturing of essential ingredients. I usually recommend Ziwi Peak, Kiwi Kitchens, and Jiminy's. Both Ziwi and Kiwi are the ideal shape, a thin, flat square or rectangle, so that each piece easily breaks into eight or more food rewards, which is essential during classical conditioning, when you might need to use several hundred food rewards over a single training session. The sheer convenience of using small pieces of dry, dividable kibble make Ziwi and Kiwi ideal as food rewards.

For lures for dogs, it's the smell, not the size, and Ziwi kibble packs a significant olfactory pop, one that is often preferred to chicken and hot dogs. (Preference tested.) Also, air-dried kibbles are, well, dry. So they are less messy for your hand, pocket, or bait bag than, say, chicken, hot dogs, or raw food.

I much prefer using kibble over commercial dog treats because commercial treats tend to be too large and too rich, and if you use them often, your dog will end up with a potbelly and the liver of a goose. That's why, during initial training, I usually recommend using kibble from the dog's daily meal ration to handfeed during training. Then stuff the leftovers into hollow chewtoys. That way, you're not increasing the dog's daily caloric intake.

That said, raw diets are wonderful for stuffing chewtoys, even if they are still messy for the floor. However, it is possible to "kibble-ize" and air-dry a raw diet to make it easier to use during training.

If you prefer canned or raw foods for your dog's everyday diet, consider switching to dried kibble temporarily while teaching basic manners, at least until you phase out food lures entirely and reduce reliance on food rewards. Then return to your preferred diet for your dog. That said, classical conditioning and progressive desensitization are ongoing to relieve fears and phobias, such as teaching your dog to love unfamiliar people, dogs, and situations. Using dry kibble is mandatory for these uses because resolution is beyond urgent, and you'll need to use so many food rewards.

Teaching Dogs to Love Kibble

At seminars, someone always tells me that their dog is not motivated by food. Well, that's why we *train* dogs. When dogs find things boring, unpleasant, or scary, we teach them that they are exciting, enjoyable, and harmless. When dogs don't like some children or men, we classically condition dogs to love them. When dogs don't like kibble, or training itself, we *teach* them that kibble and training are the best things in life. That's what training is all about: changing the perceptions and feelings and, hence, behavior of dogs.

Handfeeding Empowers Kibble as a Treat

When you handfeed kibble piece by piece, many dogs seem to enjoy the kibble like it's the tastiest treat on planet Earth. When I offer the kibble, I always pretend that I love it, saying, "Mmmmm-mmm, yum yum. Absolutely scrumptious." Then I praise the dog, "Good girl. Isn't this cool? *Chou oishii!*" Then dogs love it, too. Maybe they refine their olfactory perception because I like the kibble, thinking: *Hmm, Ian thinks it's good, maybe I should take another sniff.* The process is like decanting rotgut Scotch into a Talisker single-malt bottle. People are so influenced by the label. Say it's good, and they think it's good.

Enhanced Kibble

One way to enhance kibble is to make it smell better. Seal a couple of cups of dry kibble in a plastic bag along with a few grams of freeze-dried liver or bacon. Within an hour, the smell of the liver or bacon will permeate the nutritious kibble, so that it smells like liver or bacon. Enhanced kibble is ideal as a lure to teach Shush, as a special reward for peeing and pooping on cue, and for when children, men, and strangers handle and train your dog. Also, I like to include some enhanced kibble in regular kibble mush when stuffing chewtoys. It's kind of like the "rocks" in rocky road ice cream.

The Delinquent Waiter Routine

However, if dogs are still disinterested in kibble, I generally opt for the "delinquent waiter routine." This technique teaches dogs to finish meals, plus it desensitizes the dog's food bowl as a valued object to be protected.

Measure out your dog's dinner kibble in a bowl. Go through the usual routine that precedes feeding your dog. Then ask your dog to Sit and put the *dog's* bowl on the floor... with only a single piece of regular kibble in the center of the bowl — *an amuse bouche*. You absolutely must film this. A dog's expressions are often hilarious.

Some dogs nose-scan the bowl in disbelief, no doubt searching for the single kibble's companions and then, after shooting you a withering glance, eat the kibble. Other dogs wait a while before eating the kibble. Yet other dogs turn around in disgust, wander dejectedly to their bed, lie down, often with their back toward you, and let out a long, sorrowful sigh. But then they too eat the kibble eventually.

No matter how long this takes, the moment the dog eats the kibble, praise, pick up the bowl, and drop in *two* pieces of kibble for the next course. Repeat as many times as necessary until your dog immediately eats the kibble. Then it's time for the third course — *four* pieces of kibble in the bowl. I find dogs quickly reevaluate the value of kibble when each course comprises just a few pieces.

Obviously, you could do this just sitting in a chair and handfeeding one or more pieces of kibble from a bag. However, I thoroughly enjoy the delinquent waiter routine. It's relaxing and calms the dog, I usually end up with lots of unsolicited Sit-Stays, and it desensitizes the food bowl as a subliminal bite trigger.

To speed up the process, every time your dog licks the bowl clean, praise and slip a piece of augmented kibble or chicken in the bowl as a surprise reward, and then pick it up to "refill" with another four pieces of kibble. As Gwen Bohnenkamp, puppy trainer *extraordinaire*, used to say, "And so the dog learns . . . hands come to give and not to take."

When Dogs Blow Off the Lure

When a dog is disinterested or only moderately interested in the food lure, please don't succumb to using treats that are high in fat, sugar, salt, and preservatives. By far the most common reason for a dog to blow off a food lure is because the lure has become a bribe. Instead, teach your dog to love kibble. Or simply use another type of lure, such as a chewtoy, tennis ball, or tug toy. These toys are typically used as rewards, but they also work well as handheld lures to teach recalls, following, heeling, Sit, Down, Stand, and so on. A tug toy or tennis ball quickly brings some sparkle to even the most sluggish of heeling.

More likely, though, the primary reason why a dog is disinterested in regular healthy kibble is because they are getting better-smelling commercial treats elsewhere.

The Importance of Off-Leash Training

When I started teaching *off-leash* puppy classes in 1982, dog training quickly changed from on-leash to off-leash. Training this way, owners and puppies develop off-leash skills from the outset, and of course, there are no leashes for owners to jerk, so the focus shifts from physical corrections to motivation and using puppy play as the primary reward.

Training off-leash has so many advantages. First and foremost, that's what owner's need. Dogs live off-leash at home, indoors and outdoors. Also, by design or accident — because someone leaves the front door open, for example — at some point your dog will be off-leash outside of your property. It is vital to prepare for these eventualities.

In off-leash classes, you learn how to control your pup in a very distracting setting (with many other puppies running around). In fact, during the very first session, you must call your pup away from play, which gives you a head start before visiting dog parks. It quickly becomes second nature to frequently interrupt play by calling your pup, and then saying, "Go play," to let play resume as a powerful reward for distracted recalls. This keeps play from becoming a powerful distraction.

Off-leash classes provide puppies a wonderful opportunity to socialize with another twenty or so people — the families of the other puppies in class — and especially to meet children, which is so important if your puppy doesn't live with children. Similarly, puppies learn to get back to playing with other puppies and continue refining their social savvy after living in a doggy social vacuum for four weeks or so. Plus, this all happens in a safe, controlled setting with a trainer on the ready to tone down play styles or build confidence.

During off-leash play-fighting and play-biting, puppies learn to develop bite inhibition and tone down the force of their bites. This is so important should they, as adults, have a scrap with another dog, or growl and snap at a person.

Moreover, puppy classes are an *off-leash track* that naturally and progressively works toward training off-leash as adults. After completing Puppy 1 off-leash, puppies may continue to Puppy 2 off-leash, and then to Puppy 3, at which time they are eight-month-old adolescents with good off-leash ability.

The Problems with Starting On-Leash

On-leash training is more time-consuming because it converts a relatively simple one-step process into a longer two-step process: First train on-leash and then train off-leash, which has become an even more formidable task because the owner has developed no off-leash training skills. Also, the leash has become a crutch for the owner, and the dog has become accustomed to "listening" to the leash, rather than listening to their person.

In the nineties, a trainer from Chicago conducted what was most probably the very first dog training study, comparing the relative speed of on-leash training, off-leash (lure-reward) training, and physical prompting. The speed of acquisition of verbal requests and hand-signals was quickest off-leash, much slower on-leash, and slowest with someone holding the collar and physically prompting the required responses.

Using a leash for safety on busy sidewalks and in other public places is common sense. So it's important to train dogs to walk on-leash in distracted settings. In puppy classes, this is easily accomplished by having half the class practice off-leash stays, while the other half practices on-leash heeling and walking.

A huge drawback of on-leash training is that it often camouflages a dog's tricky temperament and does a lot to prevent resolution because the problem is "managed" by the leash. On the other hand, in off-leash classes, owners are immediately aware of any difficulties or problems and so deal with them immediately. Owners often get a rude awakening at the start of the very first class: One moment their pup is on-leash and by their side, and the next, their pup is off-leash and under a pile of puppies. Then the trainer says, "Call your puppies." Impossible? No. The owner immediately confronts the task at hand and progressively, sometimes riotously, establishes off-leash control. Indeed, by the end of the first session, many puppies Come-Sit promptly when called from play, and the owners are amazed and happy with their success.

Chapter 6

Teaching ESL

Teaching dogs the meaning of words and opening communication channels is a game changer that facilitates every aspect of our life with dogs: It allows us to explain what we would like our dog to do and how we would like them to act, to offer more detailed instructions as they learn, and to provide specific verbal guidance when they misbehave or forget their manners.

The Basic Training Sequence

As I formalized lure-reward training, I reduced the scientific principles to a single, simple four-step basic training sequence:

1. Request → 2. Lure → 3. Response → 4. Praise and reward

I wanted lure-reward training to be quick, easy-to-use, highly effective, and lots of fun for families, especially children. What could be simpler than 1-2-3-4? Of course, as soon as I wrote that, I immediately thought of: 1-2-3, 1-2, and 1.

This training sequence constitutes a combination of classical conditioning and operant conditioning: Through classical conditioning, dogs learn to love training and the trainer and to form a strong anticipatory association between the verbal cue, the movement of the lure, and the following, specific desired response. Through operant conditioning, the reward reinforces and increases the frequency of the desired response, making it more likely to occur in the future.

For example: (1) We say, "Rover, Sit," and (2) move a food lure upward and away from us, slightly above the dog's muzzle. As the dog's nose lifts up to follow the food lure, (3) the dog lowers their rear end into a sit, and so (4) we profusely praise and occasionally offer food as a reward from our nonluring hand.

After several repetitions, I imagine a dog musing: *Hmmm! When I look up as she raises the food lure over my muzzle, my butt hits the ground, and then she seems so happy. She praises me lots and occasionally gives me a piece of kibble. I love it when she's happy and food is a bonus. I love sitting. Sitting is a human happy pill. I'll become sit-happy and sit lots. But wait, when I sit on my own, she usually doesn't pay as much attention to me. Aha! I've got it! Every time she says, "Sit," a word I don't understand, she always raises her hand with the food, and then if I sit, I often get a food reward. Well, that's easy. From what I know of body language, I think her hand movement is a signal to put my butt on the ground. From now on, every time she gives that handsignal, I'm going to sit right away. I love this game and I love my person!*

Yes, the bonus of all reward-based training is that dogs learn to love training and the trainer. Reward-based training is high-octane classical conditioning!

When we start training, a dog will not understand our verbal requests, but it's still important to always give the verbal command *first*, pause briefly, and follow the request with the lure-movement. Move the lure so the dog assumes the desired position and then offer lots of praise and occasionally a reward. Comprehension is progressive. It is not as though the dog has a lightbulb moment (as I pretended above). Although the dog may not yet respond to a verbal cue, the word becomes more and more salient every time it is followed by the lure-movement.

Want to have fun with luring? Ask your dog a question like "Would you like to learn more?" Then move the food lure up and down in front of the dog's nose, and your dog will nod in agreement.

As they say with large animals, control the nose and you control the animal. See if you can move the food to lure your dog to spin clockwise. Give it a go; you'll work it out. Now try counterclockwise. Then circle around you (by passing the food lure from hand to hand behind your back), and reverse in the other direction. Try a figure eight through your legs (by changing hands with the lure twice). Now take one huge step forward and lure a figure eight and then take another step and lure another figure of eight. Now try walking with a figure eight every step. Now try figure eights when walking backward.

Here's another good one: Get your dog to back up by moving the lure under your dog's chin and toward their chest. Can you get your dog to back up in a straight line? Alternate getting your dog to back up and come forward and wow! You and your dog are almost ready to dance on *America's Got Talent*. Why not play "The Love Club" (Lorde) or "Revenge" (Pink and Eminem) — two of ZouZou's favorite dance numbers — and give it a go?

Leading your dog's nose with a food lure is the trick for luring your dog to Sit, Down, Stand, Rollover, Beg, Bow, Bang!, or assume any posture. In addition to learning the commands in this chapter, experiment with other positions and behaviors and give them a go.

Four Tips for Effective Luring

As simple as lure-reward training is, there are four key luring tips that will accelerate the speed of learning.

Don't Touch Your Dog

Work off-leash and don't touch your dog at any time until your dog responds correctly, and *then* lovingly praise and pet your dog. If you touch your dog or hold the leash prior to your dog's response, rather than focusing on the lure-movement, your dog will selectively attend to leash tugs or body contact. Dogs naturally consider bodily contact more attention-worthy than movement, and movement more relevant than words. Lucky for us, reading our body language comes easily to dogs, and they quickly learn handsignals.

Stand Still When Luring

Stand still when giving verbal requests and luring. Dogs naturally pay more attention to body movement rather than the words we say. If you move while speaking, it's as if the dog never hears what you say. Your words never make it to the conscious parts of your dog's brain.

Say, "Sit," while standing perfectly still, and when making the lure-movement, do not move your body in any other way. Then your dog will pay much more attention to what you say and to your hand making the lure-movement, since that's the only part of your body that is moving.

Use a Neutral Tone

If you change your tone for different requests, your dog will focus on your tone rather than your words. For example, many people use a louder, gruffer tone to command Down and Stay, and they use a soft, sweet tone when coaxing a dog to Come and Follow. If your dog focuses on your tone and associates it with the command, you'll always need to use the same tone every time, such as raising your voice to get your dog to lie down or acting sweet to get your dog to come. Thus, if you're in public and embarrassed about raising your voice, or you feel tired and cannot muster the enthusiasm to cajole your dog, your dog will not respond.

To check whether your dog has learned the meaning of your instructions or if they are responding to your tone, say, "Down," in a higher pitch and rising intonation and slightly raise your head, as people often do when requesting Sit. Or if your dog is lying down, shout, "Sit," in an authoritarian manner while bending over and lowering your head, as many people do when commanding Down.

Always give instructions in a neutral tone so your dog can focus on each word.

Pause Before Luring

Timing is critical for successful luring. Say, "Sit," *pause for half a second*, and then give the lure-movement or handsignal. Always speak first while standing perfectly still, pause, then lure or signal.

If you lure, or give a signal, at the same time as giving a verbal instruction, your dog will only focus on the lure-movement or handsignal. They will not attend to the words you say, and so they will not learn their meaning.

A half-second pause between a verbal request and the following lure or handsignal is ideal: That's sufficiently short for a dog to learn that your request predicts the lure-movement, yet sufficiently lengthy for the dog to respond *after* your verbal request but *before* the handsignal, which provides proof that your dog is beginning to understand the meaning of the words.

Basic Commands

When teaching Sit, Down, and Stand, rather than teaching the three body positions one at a time, I always recommend that people teach all three at the same time as part of a sequence. Why? Because we can. Although dog trainers using other techniques teach just one body position at a time, the beauty of lure-reward training is that we can teach several body positions from the outset, so why not? It's quicker.

In addition, we need to teach each of these three positions from the starting position of the other two. In truth, we aren't really teaching three body positions but actually *six* different body-position changes. Consequently, I use a very simple but extremely useful test sequence for practicing all six body-position changes equally: Sit-Down-Sit-Stand-Down-Stand, which I abbreviate as S-D-S-St-D-St.

Some body-position changes are easier to teach than others, depending on the dog's starting position. For example, Sit from Stand is much easier to teach than Sit from Down, especially for large dogs.

Similarly, Down from Sit is easy, but Down from Stand is more difficult. Often people take the "easy" route, and instead of teaching

Down from Stand, they teach the sequence Stand-Sit-Down; that is, from Stand, they first instruct Sit before instructing Down. This can get confusing for dogs, so just persevere with Down from Stand because usually, when we instruct Down, our dog is *standing, in motion, and at a distance*, for example, when playing, chasing dogs, or running away. This is why teaching Stand on cue is vitally important. How else can we practice Sit from Stand and Down from Stand unless we've taught the dog to Stand on cue?

Ultimately, we need to teach dogs Sit, Down, and Stand from *every* body position and during *every* activity.

Sit-Down-Sit-Stand-Down-Stand

To start, call your dog so that they are standing in front of you, and work through the whole sequence in order:

Sit from Stand

1. While standing still, say, "Rover, Sit," then pause for half a second.
2. With a food lure between thumb and forefinger, raise your hand, palm upward, right in front of your dog's nose and a tad over their muzzle, toward their eyes.
3. As your dog looks up to follow the lure, they will naturally sit.
4. Once they do, calmly but generously praise, and after a couple of seconds, maybe offer a food reward from your nonluring hand.

The easiest way to entice quadrupeds to sit is to lure them to look up. To understand why, stand like a dog (on straight legs, with your palms and feet on the floor) and try to look up at the ceiling. It hurts, right? Now bend your knees and squat on the floor like a sitting dog and look up. Pain-free and easy as pie.

During luring, if your dog jumps up, you've raised the food lure too high. Try to keep it within an inch of the top side of the dog's nose. If your dog backs away from you, practice with your dog backed into a corner.

If you, your family, and guests have handfed your dog on a regular basis, they will be much less excited when you lure. However, if your puppy wiggles and worries at the food, keep your hand steady, slowly but demonstratively move the food lure, and calmly praise when the dog sits. Don't move about, don't wave your arms, don't waggle the food, and don't squeak and squeal when praising. When training puppies, too much movement and excitement in your voice is like adding gasoline to their hyperactive furnace. Your dog's activity level and demeanor will reflect yours. Think of praise as a soft and gentle expression of admiration intended to mesmerize mind and body and calm and soothe your dog's very being.

Always reward your dog from your nonluring hand so your dog learns that they never get to eat the food lure, and that any food rewards come from the other hand, which makes it much easier to phase out food lures.

Sit is the most useful instruction in your Doggy Dictionary. Once your dog's response has become prompt and reliable, the simple instruction Sit prevents or resolves a very long list of behavior and training problems, first and foremost, jumping up.

Down from Sit

1. Say, "Rover, Down."
2. Hold a food lure between thumb and forefinger, and with palm downward, lower your hand from "nose to toes." Do this from right in front of your dog's nose until your palm is flat on the floor right in front of your dog's paws.
3. As your dog's nose follows the food lure to the floor, usually your dog's chest and elbows, and eventually their rump, will follow as your dog lies down.
4. Softly, meaningfully, and generously praise, and after a few seconds, maybe offer a food reward from your nonluring hand.

If your dog's nose goes to the ground but they don't lie down, try inching the food between their front paws and toward their chest to prompt the butt to slide backward. This works well on slippery floors.

Or try inching the food just a tad away from the dog's front paws. This works well on carpets.

If you're still having difficulty, sit cross-legged on the floor in front of your dog, then raise the knee closest to the dog, foot flat on the floor, to make an inverted V. Extend your luring hand under your raised knee and up to your dog's nose. After saying, "Down," lower your hand, palm to the floor, and draw it back through your knee-tunnel, away from the dog. To follow the food lure, the dog must lower their withers to get under your leg. I know, it's a bit of a rigmarole, but this is fun, right? Luring dogs and Pilates at the same time.

If none of the above works, use a training table. Think lions, tigers, seals, and elephants. This is a two-foot-square table with eight-inch legs. (We use Blue-9 KLIMB dog training platforms; expensive, but stable and worth the money.) The same one-size-fits-all table may be used with puppies and huge dogs. With your dog sitting and their front paws within an inch or two of the table's edge, lower the food lure, palm down, just below the top of the table. As your dog lowers their nose and forequarters an inch below the tabletop, they will have to flatten their rear end as ballast to stop from tipping forward.

First, make sure to familiarize your dog with the table by luring "jump up" and "jump down," and preferentially rewarding your dog each time they jump (or climb) onto the table.

Sit from Down

1. Say, "Rover, Sit."
2. Hold a food lure between thumb and forefinger, and with palm upward, raise your hand from right in front of your dog's nose to where you expect the dog's nose will be once sitting again.
3. As your dog follows the lure, they will sit up.
4. Praise and, after a few seconds, maybe offer a food reward from your nonluring hand.

From Down, most small and medium-sized dogs immediately spring up into a Sit. But large dogs? Some are down for the count, and so you'll probably need to energize them. Luring will need a little more energy on your part. Say, "Sit," and then quickly take one step back, stand tall, and waggle the lure in front of your dog. The instant your dog energizes and starts to stand up, step in quickly and lure Sit.

Stand from Sit

1. Say, "Rover, Stand."
2. Hold a food lure between thumb and forefinger. With the palm facing your dog's nose, move your hand away from the dog.
3. As the dog's nose follows the lure, they will naturally stand.
4. Praise and, after a few seconds, maybe offer a food reward from your nonluring hand.

Some dogs stand but then immediately resume sitting. Well, at least you know that your dog loves sitting, but to prevent this, use a two-part lure-movement. Move the palm of your hand away from your dog's nose, and as soon as your dog's butt leaves the floor, lower your hand just one or two inches, slightly under the dog's muzzle, to get your dog to look down just a little and arch their back to prevent them from sitting again. If your dog lies down, you lowered the lure too much; just repeat the procedure after getting your dog to Stand from Down.

Stand-Stay is a very useful instruction for grooming and veterinary examinations. Take as much time as necessary to solidify your dog's Stand-Stay and progressively increase duration. As soon as your dog stands, keep hold of the food lure to keep the dog's neck arched for longer and longer before offering food as a reward from your nonluring hand.

Down from Stand

1. Say, "Rover, Down."
2. Hold a food lure between thumb and forefinger, and with palm downward, lower your hand from "nose to toes" until your palm is flat on the floor.

3. Most dogs readily lower their nose to the floor, and eventually, their butt goes down.
4. Praise and, after a few seconds, maybe offer a food reward from your nonluring hand.

Ah, but what if your dog lowers their nose and elbows to the floor but leaves their butt sticking up in the air, perfectly comfortable in a playbow? You pretty much make the same adjustments as when teaching Down from Sit: Either slowly move the food lure between the front legs and toward the chest to prompt the butt to slide backward; try the raised-knee tunnel; or work on a training table. These days, I teach nearly all body-position changes on a training table as the default. It's so quick, and there's no need for a knee-tunnel, canine version of Twister.

However, you could also change to Plan B and teach your dog Playbow on cue. This is too good an opportunity to miss. Praise your dog, "Good playbow. Gooood playbow." Then keep the dog in a playbow for as long as possible. Then alternate Stand and Playbow until the dog has fluency in both cued positions before resuming to teach Down from Stand.

Want a laugh? Go back to Playbow and alternate Playbow and Down. It's pretty easy to instruct a dog to Playbow from Stand and then lie down on cue, but to do the reverse and instruct Playbow from a Down-Stay is more difficult. But stick with it. One of my all-time favorite owners in class taught her darling Irish terrier, Fenor, to Playbow from Down, and the entire class looked on in awe as Fenor's butt slowly rose to a Playbow from his perfect sphinx position.

Stand from Down

1. Say, "Rover, Stand."
2. Hold a food lure between thumb and forefinger, and with the palm facing your dog's nose, move your hand upward and away from the dog.
3. As the dog's nose follows the lure, they will naturally stand.

4. Praise and, after a few seconds, maybe offer the food as a reward from your nonluring hand.

If your dog doesn't stand, energize them first, and if your dog stands but then immediately sits or lies down again, use the two-part lure-movement, as in Sit from Down.

Off and Take It

A common problem with luring is that the dog gets *too interested* in the lure and starts to worry at the food, nibbling, licking, mouthing, biting, slobbering, and pawing to get at it. If this becomes too much of an annoyance, use a wait-and-reward training technique to teach Off and Take it.

The verbal instruction Off means don't touch what you're trying to touch, whether food in the hand, food on the counter, another dog, a baby, or the neighbor's cat. It's an instruction with a hundred and one uses. With a new puppy or dog in the house, Off will probably be the second-most-frequently-used request after Sit.

Hold a single piece of kibble in your fist right in front of your dog's nose. Then wait and watch. Give no instructions whatsoever, just *keep your hand perfectly still.* Your dog will attempt all sorts of ploys to extricate the food from your hand. *Ignore them all.* Don't get frustrated if this takes a minute or two. In fact, the longer the dog worries at the food, the better they learn that these tactics don't work.

Eventually, your dog will pull their nose away for *an instant.* The exact moment your dog ceases contact with your hand, immediately praise, "Good dog," and open your hand so that the dog can take the food from your palm. Do this again, but this time wait for half a second of noncontact before praising, and then offering the food. Next, go for a full second of noncontact. Then two seconds, three, five, eight, twelve, and fifteen. I love to count out the seconds in "good dogs": "Good dog one, good dog two, good dog three..." Or mesmerize the dog with praise when not touching your hand: "Gooooood dog, therrrre's a goooood dog."

This technique is simply amazing and prevents so much frustration. When teaching Off, we don't give any instructions. The only words we use are "good dog." But dogs quickly learn from the consequences of their actions: Keep touching the food and you'll never get it; don't touch the food and you'll get it shortly. Additionally, if you hold the food just a little higher than your dog's nose, your dog will give you lots of free Sit-Stays as they look up at the food in your hand. Teaching Off is a marvelous impulse-control exercise that teaches dogs to "get a grip" on their excitement, wiggle less, and focus more.

Now let's add some instructions. Before presenting the food in your fist, say, "Off." Maybe move your hand from side to side, up and down, or away and toward your dog to tempt them just a little. If at any time your dog touches your hand with muzzle or paw, repeat, "Off," and then restart the count as soon as your dog ceases contact. Once you're satisfied, say, "Take it," and open your hand to let your dog take the food. Keep track of your dog's best times as a benchmark to surpass the next time you train (such as during the next hourly potty-play-training break). Once your dog gets the idea, present the food between thumb and two fingers, as you would when luring.

Then, when you go back to luring practice, instruct Off any time the dog touches your hand, and they will understand what you want. When luring, you want the dog to follow the food in your hand, but not touch.

When teaching Off in this fashion, praise lots. The only time your dog requires guidance is if they hurt you by scraping your hand too vigorously with their nails or biting at your hand too hard. Respond by saying, "Ow! That hurt, you worm!" Don't shout, but let your dog know they hurt you. Maybe take a short time-out to let your dog calm down before trying again.

Dogs worry at food in our hand because they want it, and when we move our hand, that makes the food even more exciting. This is the reason to keep your hand perfectly still until your dog eventually stops worrying at the food.

When filming my UK television series, one lady had multiple scars on her arms from her German shepherd puppy, Shogun. It took me only one or two minutes to teach him Off and Take it. But whenever she tried, her puppy "mauled" her hands and arms. She simply could not keep her hand still. Instead, she would snatch her hand away from the dog's muzzle, which of course was enticing her dog to grab it. (This is the most common way for teaching protection dogs to bite!) She couldn't even keep her hand still in front of my face when I pretended to be her shepherd. In the end, I held her wrist steady and bingo! Shogun learned in a flash. After a couple of Ian-assisted Offs, at last she could do it on her own.

Teaching Off and Take it greatly facilitates phasing out food lures because it teaches dogs impulse control around food, and it makes dogs calmer and more attentive.

Phasing Out Food Lures

Food lures are extraordinarily effective for teaching ESL. By focusing on the movement of our hand and arm, dogs learn handsignals very quickly. However, food lures *must* be phased out within the first few sessions. It will be easier to phase out food lures if you've been rewarding from your nonluring hand, so your dog has already learned that they never get to eat the food lure, but they might receive a reward from the other hand.

Once you have luring proficiency and can lure at least twenty randomized responses in a row, and each with a single lure-movement, it's time to have a go at using empty-handed handsignals. This is much easier than it sounds. And before you gasp, "Twenty in a row!?!" I should mention that during week one in puppy classes, the average golden or Lab pup will happily perform twenty consecutive lured Down-Sits — aka, Puppy Push-Ups. *That's forty behaviors for the prospect of a single treat.* So, I repeat, once you have luring proficiency, dispense with food lures.

Of course, it will be necessary to use food lures again when teaching any new exercise. It's also OK to return to food lures if your dog needs an occasional refresher. Also, it's OK to continue to use food as an occasional *reward* even after you've stopped using it as a lure.

Put the Food in Your Pocket and Go for It!

Practice luring the S-D-S-St-D-St sequence over and over and over, using verbal requests followed by the appropriate food lure-movement. *These six position changes are the building blocks of all training.*

Practice several sequences in succession. After the final Stand in the sequence, instruct Sit to begin the next sequence. Maybe practice multiple sequences while sitting in an armchair. Use half of your dog's dinner kibble from a single meal to practice luring multiple test sequences.

During every sequence, a dog is progressively learning that the verbal request predicts the lure-movement, and that the lure-movement predicts *possible* food delivery *after* the correct response. Only reward your dog for the better (quicker and snazzier) position changes. Be discerning. Do not reward your dog after every response (for more on why, see chapter 8).

Now comes the big moment. The process is so simple: *Put the food in your pocket or bait bag* and give it a go. Step back and call your dog, say, "Rover, Sit," and then quickly give a flourishing, demonstrative, *empty-handed* handsignal to Sit — *just as if you had food in your hand.* When your dog sits, praise and give your dog a food reward from your other hand. Repeat Come-Sit a few times until it becomes automatic for your dog. Now move on to practice Come-Sit-Down. Then Come-Sit-Down-Sit, and Come-Sit-Down-Sit-Stand. Soon you'll be able to run through the entire S-D-S-St-D-St sequence *without the need for food in your hand.*

Try three S-D-S-St-D-St sequences in a row. This is the basic test I give to all dogs. I do this every step of the way when training to monitor progress, each day, and in each session.

Congratulations! You have just conquered the most difficult part of lure-reward training. You have gone cold turkey on food lures! I mean it. This is where so many, many people stall. They continue luring with food until food lures become bribes, which leads their dog to only comply occasionally and according to whim. From now on, your training will literally whoooshhh!

Celebrate your and your dog's success. Repeat the entire sequence once more with you beaming with pride and your dog receiving a literal bunch of food rewards. This is special. Don't worry about how many food rewards you give your dog. It's OK to occasionally go over the top. Once you start the motivational stage of training (the real fun part), reducing the number of food rewards is easy.

Troubleshooting Problems When Phasing Out Food Lures

Here are a few tips should you encounter hiccups. Always work as quickly as possible to keep your dog engaged, and use exaggerated, confident handsignals. Confidence is key. If your dog does not respond immediately to the first empty-handed handsignal, then repeat the handsignal, and do this multiple times if necessary. It's totally OK to repeat the handsignal. Your dog is still learning, and comprehension is progressive. Many times, I've seen a dog fail to comply on the first attempt at an empty-handed handsignal, but they do so on the second attempt. Or they fail to comply on the first four attempts but do so on the fifth.

If you've given five handsignals with alacrity and gotten no response, quickly put your hand in your pocket, grab a piece of kibble, and food lure just one response. Then return to giving an empty-handed handsignal for the next response. Here is a good trick I like to use: If I've gotten no response after five empty-handed handsignals, I quickly dip my hand in my pocket (as if to get a piece of kibble) and then whip out my hand and give an expansive, blossoming, but *empty-handed* handsignal. Works every time.

The biggest problems with phasing out food lures is that they work like magic, and people simply *don't want* to give them up. Also, they fear that without them, they'll lose their control over their dog. In essence, food-in-the-hand can become a "mental crutch" for the trainer, and so when they try to work without lures, their verbal requests and handsignals become unsure, tentative interrogatives — *Will you sit for Daddy, please?* It's not so much that dogs are insensitive to pleading, but rather, dogs often don't recognize or comprehend the instruction because of the weird, altered, uncertain tone.

Once you've phased out food lures, you'll be able to get equally rapid position changes with handsignals because they are so easy for the dog to follow. Just think how impressive it would be if you repeated three consecutive S-D-S-St-D-St sequences with handsignals in a dog park. Your dog would be the center of attention, and everyone would come and ask how you did it.

Meet the Beast: Working without Food on Your Person

If you are still having difficulties phasing out food lures, and your dog is uncontrollably wiggling and jumping, I would take a short training break to get your dog to calm down and focus and get used to you working with no food on your person. Let's "meet the beast" and solve this problem once and for all. Using this wait-and-reward technique, you can use the distraction of kibble to teach your dog to overcome that distraction and focus on you.

Let your dog watch you measure out a meal's worth of kibble in a plastic bowl on a table in the center of the room, or on a countertop, so that your dog knows it's there, but it is too high to reach. Take two steps back (behind your dog) and wait and watch.

No doubt your dog will be riveted on the food, looking up at the bowl and intensely sniffing the odor-waterfall spilling over the counter's edge. It won't take long. Usually within fifteen seconds, your dog will take a quick glance back at you for maybe 0.2 seconds, as if to inquire whether you have any suggestions for resolving this significant dilemma: *Food too high, me too short.* This is the behavior you're looking for. The instant your dog glances at you, praise, quickly step forward, pick up a piece of kibble, give it to the dog, and then step back behind your dog again.

This time wait for a palpable half-second glance at you before praising, stepping forward, and offering a piece of kibble. Then wait for a full second stare, and then two seconds, three, five, eight, ten, and so on. Keep prolonging the duration of riveted attention necessary for you to step forward to reward your dog with kibble — as your dog pleads, *Hey, could you help me here, I can't reach the food?*

Now step back a little further each time and wait for your dog to look at you and approach a little before praising and going back to the bowl to offer kibble. Then wait for your dog to come all the way to you. Eventually, you'll reach a point where the gravitational pull of the bowl of kibble is not as powerful as the obvious key to kibble access: approaching and looking at your face. But it gets even better. Your dog will inevitably approach all the way and choose to sit right in front of you, since of course, sitting makes it so much easier to look up at your face when close, and especially if you stand ramrod straight. Now each time after you walk away from the bowl, wait for your dog to Come-Sit-Stay-Watch before praising and returning to the bowl to give your dog a food reward. In successive trials, progressively increase the required duration of the Sit-Stay-Watch.

If your dog runs circles round you, paws, or jumps up, just ignore your dog's silly antics, the same as when teaching Off, and wait for a solid Sit-Stay-Watch.

Now start walking around the room. You'll probably find you have company. Ah yes, your dog is walking by your side, and whenever you stop and stand still, your dog will sit in front of you and gaze into your eyes. Praise your dog. No need for exquisite timing; just periodically praise or gently stroke your dog and on occasion, say, "Hey, good dog, that's worth a piece of kibble," and walk back to the bowl. Your dog will no doubt accompany you and immediately Sit-Stay-Watch the moment you stop. If not, stand still and wait for it.

After each reward from the bowl, off you go again, walking around the room. Make lots of turns and change speed frequently; your dog will most likely follow and automatically Sit-Stay-Watch anytime you stop. When that happens, praise lots and fill up their affection tanks.

Let's consider what you're doing. You're working your dog off-leash *without any food on your person*. You've successfully, completely, and utterly phased out the need for food lures to get a Come-Sit-Stay-Watch and a pretty good Heel with automatic Sits.

Now, use your dog's attention to resume lure-reward training again, but without any lures. Ask for what you want using only verbal requests and handsignals. Each time I stop walking, I like to use the Robert De Niro "I'm watching you" handsignals — two fingers pointed at my eyes and then an index finger pointing at my dog's eyes. Makes people laugh when I walk my dogs.

If you want, use this opportunity to teach a new cued behavior, such as Beg (Sit Pretty) or Rollover, using only your hand as a lure. Try to work out how to do the hand movement yourself from what you've learned already. Or check out the books in the resources.

A variation of this exercise is to enlist the help of an assistant. Together, you tag-team your dog. One person gives all the instructions and praise, but a second person acts as the National Bank of Food Rewards. When the dog performs well, the first person praises lots, but only occasionally says a designated phrase — for example, "Goooood girl" — that signals the "Bank" to step forward to reward the dog.

From now on, whenever you feel overwhelmed during training, take a wait-and-reward break to relax, slow your heart rate, and calm your brain and your dog. Take half a dozen pieces of kibble, sit in a comfy chair, and wait and watch as it all unfolds before your eyes.

Phasing Out the *Necessity* of Handsignals

Once you can work your dog without food in your hand as a lure, training will become plain sailing. Full speed ahead! Once your dog understands the meaning of handsignals, the basic training sequence becomes:

1. Request → 2. Handsignal → 3. Response → 4. Praise, maybe reward

Handsignals have so many uses, so we would never want to phase them out entirely. For example, we can use them to signal our dog to Shush or Settle while we are chatting to someone without having to disrupt the conversation. Additionally, people are so impressed when a dog performs a position-change routine or heeling with handsignals.

In 1982, after giving a talk about off-leash puppy classes, I was approached by Suzi Bluford, who said, "With respect, Dr. Dunbar,

I simply cannot believe what you said." At the time, Suzi's golden retriever, Streaker, was the number-one Obedience Trial Champion in the United States. I invited Suzi to observe classes, and while other dogs were spread out practicing stays and following, I picked a golden puppy, and using handsignals, I put him through the position-change portion of the AKC Utility Signal Exercise: Heel-Sit-Heel-Sit-Heel-Stand-Stay-(walk away)-Down-Sit-Come-Sit. Her jaw dropped. After class, she said, "That takes months to train! I want to do that!" Suzi became my second trainer for SIRIUS, and she taught puppy classes for thirty-five years. My first trainer, Laura Enos, is the only puppy trainer who has taught puppy classes continuously for longer — forty-one years and counting.

Nonetheless, we don't want to rely entirely on handsignals. We must confirm that dogs respond to verbal instructions, so that we have control when our dog is out of sight, or cannot see us, such as when their back is turned or when in a different room. Consequently, we must check that handsignals are not necessary and that verbal requests work without them.

Phasing out the need for handsignals is much, *much* easier than phasing out food lures. Simply repeat the S-D-S-St-D-St sequence, but with a longer, one-second pause between the verbal request and the following handsignal. The association between the two has already been forged because, right from the start, we have always given a verbal request before luring/signaling. We just need to give the dog a little extra time to respond to our verbal cue. Look for when your dog responds *after* your verbal request but *before* the handsignal, praise enthusiastically, and offer several rewards. Once handsignals become unnecessary, the basic training sequence becomes:

1. Request → 2. Response → 3. Praise, maybe reward

Teaching Sit-Stay, Down-Stay, and Stand-Stay

In the above "meet the beast" wait-and-reward exercise, we teach the dog that "Come" means Come and Sit and Watch, and that whenever we stop walking and stand still, that means Sit. Now let's teach dogs that Sit, Down, and Stand mean Sit-Stay, Down-Stay, and Stand-Stay, which makes the instruction Stay unnecessary.

While repeating the S-D-S-St-D-St sequence, have your dog *remain in position* for a few seconds after each position change simply by praising, *but delay* reaching for the food reward from your pocket or the kibble bowl. Once you are no longer using food as a lure, you may reward your dog from either hand. Once you have the food reward in your hand, further delay giving it to your dog, so that your dog doesn't develop an eat-and-run habit. Instead, hold on to the food reward and instruct Off. *The longer you hold on to the kibble, the longer your dog will stay.* With each repetition of the sequence, gradually and progressively increase the duration of the short *stay-delays* from a couple of seconds to three, five, eight, and ten seconds.

Solidifying Stays

As stays increase in length, praise your dog intermittently throughout (until released by the next command). Every few seconds, praise for a solid stay, give better praise for better attention and stability, and as much calm but heartfelt praise as you can muster for concrete-elf stays. In a surprisingly short time, you will build minute-long Sit-Stays and Down-Stays and half-minute Stand-Stays.

Practice Stand-Stay as much as possible. Why? They are usually *shorter* because they are less stable. This means that practicing a one-minute Stand-Stay is the equivalent of practicing multi-minute Sit- and Down-Stays. In a sense, practicing Stand-Stays is the quickest way to get very lengthy Sit- and Down-Stays. Each session, make note of your dog's longest stays and post it on the refrigerator door with a challenge for other family members to try to do better.

At first, expect that your dog will feel an urge to break a stay on their own. Watch closely for the signs that this is about to happen and turn it into a learning experience. If your dog repositions in a Sit-Stay, or your dog's eyeballs move a fraction, immediately reinstruct, "Sit."

This is how a dog learns that Sit means "sit *still* and *look at me* until I give another command."

Here's the deal: *If a dog looks away or sniffs away, they will soon go away.* So if your dog's nose twitches, or if the eyeballs even bobble, way before your dog's attention starts to wander, immediately reinstruct Sit (or repeat the command for the appropriate position), *before your dog starts to wander*. Then praise your dog once the stay becomes solid and attentive once more.

Even if your dog breaks the stay and moves out of place, say, "Sit." If you switch commands, such as to Come or to another position (like Down), or physically lead the dog back to where they were, that will confuse them. The goal is to teach your dog to understand that Sit means "Sit-Stay until released."

Typically, dogs break stays and move away because their person's attention wandered. It's *our* lack of attention that allows our dog to wander off. Next time, catch the warning signs earlier. Most dogs give ample warnings that they are about to break a stay. Notice any indication, like a momentary eye flicker, that seems to say, *I'm becoming distracted and going to end this. I have other things to do.* Then *immediately* reinstruct. Some dogs, such as Malinois and border collies, can break in a flash, so you must be vigilant. Conversely, once your dog understands that Sit means "immediately sit and remain sitting *until another instruction*," you will find that they will win any staring contest. Border collies were bred to "have the eye."

Solidifying Stand-Stays

To stabilize Stand-Stays, use positive *thigmotaxis*, also known as the *oppositional reflex*. This is the impulse to push against a push or tug against a tug. This reflex is why using leash corrections for dogs that pull often encourages pulling. Indeed, most horse and stock people quickly learn that pushing on an animal's rump usually activates all the muscles that resist the pressure. I learned this the hard way as a youngster. I tried to squeeze between a dairy cow and a wall, and I pushed on the cow's side to get more room. Yup! She moved one hind leg toward me, trod on my foot, and pinned me against the wall for several minutes.

With your dog in Stand-Stay, hold a food lure about an inch in front of and slightly lower than your dog's nose. Instruct, "Off," so your dog doesn't worry the food. Then place your hand on your dog's shoulders and gently and rhythmically push, alternating between gentle pressure and release. This firms up the shoulders and front legs. Then gradually move your hand down the dog's spine while doing the same, until you're gently and rhythmically pressurizing the dog's rump to firm up the dog's rear legs and hindquarters. If your dog sits, just reinstruct Stand, and press a little gentler next time.

Releasing a Dog from a Stay

To build attentive, rock-solid stays, you must clearly communicate their onset *and offset*. Obviously, the appropriate position request (Sit, Down, or Stand) is the cue for the dog to move into position and remain in that position. Additionally, though, you must clearly and explicitly release your dog from each stay. Do this either by giving another instruction — such as another request in a position-change sequence — or by providing a *release command*, such as "Free dog."

Omaha's release command was "Hang loose," and I'd either say the phrase or give the hang-loose gesture. Being British, I knew I would never say that phrase in everyday conversation, and my hand would never involuntarily make that gesture. Another release command I really liked was used by an ex-Marine, who would say to his dog, "Dismissed," and salute as a handsignal.

It is very important to choose a release command that you would never say in common parlance. The daftest release command I have ever heard, yet is used by a surprising number of people, is "Good dog." With this as a release command, it becomes impossible to praise a dog while they remain in a stay.

Now, if you ever need to repeat a release command... woooo! There's proof for how difficult it is for dogs to learn the meaning of words. If there were ever an instruction that all dogs should instantly "obey," it would be a release command. Unless, of course, they have already grown to love training so much that they don't want it to end. Similarly, when teaching Chow, an instruction that communicates that a Sit-Stay is over and a dog may begin their dinner, with a duck-snapping-its-bill handsignal, we should expect an immediate response. This is why we frequently check a dog's comprehension of all the verbal requests and handsignals as we teach them.

Clear release commands are so important. Otherwise, dogs will improvise and learn to release themselves.

I remember an old shepherd from Yorkshire who came to one of my workshops in England. After practicing stays, we broke for lunch. While the other participants ate, the shepherd — in his tweed flat cap, trousers, waistcoat, and worn white shirt with no collar — casually leaned against a tree, rolled cigarettes, and smoked. As he did, his dog sat, virtually vibrating, his eyes piercing the shepherd's, *for the entire lunch break*. After people straggled back to the field, I said, "Let's begin." The shepherd casually pinched out his ciggie, put the butt in a tin, and mumbled one word to his dog, who *exploded* — backing up excitedly, running back to him, playbowing, twirling and twisting, while *still not taking his eyes off the shepherd*. The shepherd then nonchalantly sauntered back to join us, with his vigilant collie still on optic alert. Awesome. I think he was making a Yorkshire statement.

Come-Sit and Heel-Sit

In addition to the position-change sequences that are the basic building blocks of all training, practice repetitive, one-step Come-Sit and Heel-Sit sequences. The longer a recall or heeling sequence, the more likely your dog will become distracted and go off track, and so reduce both to the simplest and most important format — the *final step* before an automatic Sit. Think of heeling as very short Sit-Heel-Sit sequences and recalls as very short Sit-Come-Sit sequences. As Sit-Stays become stable, the dodgy part is when a dog is off-leash and in motion. So start practicing with single-step Come-Sits and Heel-Sits.

Walk backward across the room, one step at a time, practicing repetitive one-step Come-Sits, and then without turning around, return across the room to practice repetitive one-step Heel-Sits.

Short Repetitive Come-Sits

To teach your dog that Come means Come-and-Sit-Stay-Watch, start with your dog sitting in front of you and looking up at your face:

1. Say, "Rover, Come."
2. Take just one step backward, open your arms wide, and with a big smile, make noises that encourage your dog to approach.
3. As soon as your dog starts to approach, praise, quickly instruct, "Sit," and praise again.

Repeat one-step Come-Sits over and over. Occasionally reward your dog for quicker recalls and sits. Then practice two-step Come-Sits, then three-, five-, eight-, and ten-step Come-Sits, and so on. If ever you feel that you are losing your dog's attention, immediately instruct Sit. The dog's Sit-Stay-Watch is the time to take a breather and recover composure.

For longer recalls, especially when your dog is incoming fast, instruct Sit when your dog is two dog-lengths from impact.

Short Repetitive Heel-Sits

To teach your dog that Heel means Heel-and-Sit-Stay-Watch, use a food lure in both hands. Start with your dog sitting, facing forward at your left-hand side.

1. Say, "Rover, Heel."
2. Move the lure in your *left hand* from left-to-right in front of the dog's nose and then up to your belly and take one enormous, exaggerated step forward.
3. As your dog moves to follow, say, "Sit," and with the lure in your *right hand*, reach across the front of your body and lure

your dog to sit by your side. When your dog sits and looks up at you, praise and occasionally reward.

Repeat numerous, repetitive, one-step Heel-Sit sequences, and then gradually and progressively increase the number of heeling steps: Take two steps, then three, five, ten, and so on. There ya go, you're heeling your dog. Again, if you ever think you are losing your dog's attention, immediately say, "Sit."

You will find leash-walking much easier if you first teach your dog to heel *off-leash* at home — indoors and in your yard — *before* attempting your first walk.

Preventing Anticipation

The S-D-S-St-D-St test sequence is to ensure that you are practicing all position changes equally. However, if you keep repeating the same sequence over and over, quick-learning dogs, such as border collies, Malinois, and papillons, will start to anticipate what comes next.

It is indeed wonderful when a dog is an eager beaver, but we want to teach our dog to focus, listen to our voice, watch our handsignals, and follow *our* instructions rather than improvising. Anticipating the next command can be a disaster in obedience competitions, as well as in real life, especially if a dog anticipates a release command from a stay.

One way to avoid anticipation is to *randomize Sit, Down, and Stand with variable-length stays in each position.* This prevents the dog from anticipating *which* instruction comes next and *when* it is coming.

For example, the sequence might be: Come-Sit-Come-Down-Heel-Stand-Sit-Down-Sit-Heel-Stand-Down. When practicing with your dog, *always randomize a minimum of three behaviors at a time and vary the duration of each.*

Once you start randomizing, your dog's name recognition and attention improve because, after you say your dog's name, they must pay attention and listen to find out which body position is requested.

Should your dog still anticipate the next command, or move out of their current position, well, ten points for enthusiasm, but zero points for guessing incorrectly. Don't scold your dog for eagerness; simply reinstruct the position your dog should be in. For example, if your dog is standing and you instruct Down, but your dog sits instead, reinstruct Down, and keep repeating Down until your dog lies down. Trust me, this is the quickest way for a dog to learn.

Expanding Vocabulary: Fetch and Go To

Of course, the verbal requests in this chapter are only a start. Plus, there are all sorts of everyday words you can teach your dog to expand their vocabulary. Two of my favorite commands as vocabulary builders are Fetch and Go to. Although teaching Fetch was once a problem for some obedience and working competitors, from looking at the dogs in many dog parks, teaching Fetch appears to be no problem for most owners.

Fetch is useful for teaching the names for objects. As in, fetch *what*? So teach your dog to fetch chewtoys, balls, a Frisbee, a tug, your car keys, and their leash. Phoenix even learned to "fetch Mittens" (the cat).

Meanwhile, Go to is useful for teaching the names of places and people. For example, teach your dog to go to their bed, outside, inside, the car, the back seat, upstairs, downstairs, and the couch. To teach the names of people, such as family members and friends, use Round-Robin Recalls.

Round-Robin Recalls

Imagine we have two parents — Mike and Amanda — a daughter, Savannah, and a dog named Dude. Mike instructs Dude to Come (and Sit) and then says, "Dude, Go to Savannah." Mike should say this *just once* and not repeat the instruction. Otherwise, Dude will listen to him and ignore Savannah. As soon as anyone says, "Go to," they relinquish control over the dog, and control is passed to the named person.

In this case, Savannah immediately says, "Dude, Come," and she opens her arms wide as a welcoming signal. As Dude arrives, Savannah says, "Dude, Sit," and signals Sit if necessary. Once Dude sits, Savannah takes hold of Dude's collar with one hand and praises Dude while offering a food reward with the other.

Now Savannah chooses where to send Dude. She says, "Dude, Go to Amanda," and Amanda immediately calls for Dude. Then Amanda says, "Dude, Go to Mike," and then Mike sends Dude back to Amanda. Keep going like this, making sure everyone sends the dog to everyone else, so that the dog eventually learns that no matter who says a name, such as "Savannah," Savannah always calls.

Down-Stays

There are many positions for a Down-Stay. These are fun to teach, and they all have their uses.

- **Sphinx:** A magnificent, short duration, *working prone-down*. The front legs are forward, and hind legs are tucked under the rear, ready to spring into action.
- **Down:** A good-looking, longer duration, more stable, *"obedience" prone-down*. The front legs are forward, the rear is rolled onto one haunch, and both hind legs are out to the side.
- **Snooze:** A long duration, *restaurant or bar prone-down*. The chin is resting on front paws, the rear is on one haunch, and rear legs are to the side but with paws tucked under the body.
- **Settle:** A much longer duration, *"I'm working on my computer" side-down*. All four legs are stuck out the same side.
- **Bang!:** A stable, short duration, *undercarriage-examination supine-down*. All four legs are stuck up in the air with front paws flopped over.
- **Curl:** An extreme duration, *comfy couch-down*. The nose and face are hidden under the tail, as if the dog is trying to be invisible.
- **Captain's down:** An interminable duration, *"I'm the captain" supine-down*. The front legs are extended in the air and in front of the head, and hind legs are extended, so that Ian and Gina are sitting seven feet apart on our specially purchased, nine-foot, leather dog couch.

Chapter 7

Measuring Success

When I started giving doggy lectures in the seventies, most attendees were professional dog people, and most competed in obedience, so they were used to judges objectively testing for performance reliability. They knew the consequences of occasional "glitches" in performance could be significant. Just one broken Sit-Stay could mean getting a zero on an exercise and a nonqualifying score on the whole trial, which meant the combined cost of transportation, hotel stay, hotel food, and time off work was all down the drain.

I didn't have to convince my attendees that testing for reliability was critical to successful training, but these days, while most seminar attendees are still dog trainers, very few compete in obedience or working trials. So I find myself having to explain all over again why I feel so strongly that even the average person training their household companion should regularly check for "proof of training" by quantifying the ongoing changes in a dog's response reliability to cued instructions. This allows us to measure the speed of training and, hence, provides a means to monitor and assess training.

Every step of the way, we must confirm that training is indeed working — that we are changing behavior in the intended direction by some measurable criterion. To do this, I regularly use two very simple but extremely revealing indices: response-reliability percentages and response:reward ratios.

Response-reliability percentages provide a single precise score of both the dog's comprehension of the cues we use, that is, verbal requests and/or handsignals, and the level of motivation to respond. Regularly testing for comprehension is vital because lack of understanding of verbal instructions is by far the most common reason for misbehavior and noncompliance.

Response:reward ratios enable us to track how well we are decreasing our dependence on food rewards.

Repeated testing allows owners to appreciate how control over a dog can vary greatly from scenario to scenario, place to place, person to person, when on-leash and off-leash, and when around people, other dogs, other animals, and any other distraction.

Moreover, nothing is more motivating for owners than to see palpable proof of progress.

Baby Steps Are Motivating

Every time I explain the importance of quantifying behavior, I am reminded of a distraught owner of a Jack Russell terrier puppy. I noticed her sitting in a chair and sobbing after class. I hustled everyone out and gently asked her what was wrong. She couldn't answer for quite a while, then between sobs, she blurted out, "It's him!" She pointed to her puppy. "I never wanted a dog, let alone a puppy, but my kids clamored for one, and my husband chose a Jack Russell because of Eddie on *Frasier*, and..." Her tale was interrupted by more sobs. "He... he... *won't sit still for a second*, and I work all day and have to cook and clean in the evenings... and..."

Ironically, given her puppy's problem, he remained sitting calmly and gazing up at her with concern as she tried to talk. I was so tempted to say, "If you want him to Sit-Stay, just sit on a chair and cry," but clearly, this was no time for humor. (However, jokes apart, sitting and crying *is* a brilliant training technique to get your dog's attention if you're at your wit's end because your dog won't listen or respond.)

I remembered that the entire family came for the first week of class. On the second week, it was just her and the kids, and the third week, this one, she was alone. That is unusual for puppy classes because we always implore the entire family to attend every session.

After a while, she stopped crying and ever so sweetly scratched her pup's ears, which almost made me cry. I said, "Look, I can help you. I grew up with a Jack Russell and I just love them. They remind me of my uncle." That's actually true: My uncle had a Jack Russell on the farm when I grew up, and when Scampi died, he never got another dog, even though he lived to a hundred. However, no matter what breed or type someone has, I typically just insert that breed in my uncle's story. It's strange, but I find that people are more likely to listen to you, and assume you know what you're talking about, if you've trained or even known an individual dog of "their breed."

I said, "If you can stay a little while, we can teach him a Sit-Stay." So I asked the little tyke to Sit. And he did! For precisely 0.2 of a second. She was correct; he didn't sit still for a second. How did I know so exactly? Because I always use a stopwatch. As you may notice, here and throughout, I rarely use adjectives to define duration.

The woman was unimpressed. The second time I asked, the dog sat for the same length; the third time, he sat for 0.6 seconds; and then on the fourth try, he sat for 1.2 seconds!

I showed her the stopwatch, but she remained unimpressed, and so I asked her, "Look, if I were to give you a stock tip that's a shoo-in to increase by 600 percent overnight, how much money would you invest?" She replied, "The house!" I said, "Precisely! Your puppy's Sit-Stay just improved by 600 percent in just four tries, and in less than thirty seconds."

When this didn't move her, I continued: "Do you realize that if your Jack Russell terrier continues at that rate of learning and gives us 600 percent improvement just *three* more times, you'll have nearly a four-and-a-half-minute Sit-Stay!"

She looked quizzical and a bit interested but was still far from believing. I kept training her dog, and after about five minutes, he produced a ten-second attentive Sit-Stay, and after twenty-five minutes, he broke his personal best and the one-minute barrier: a one-minute-and-seven-second Sit-Stay. That was a 33,500-percent improvement!

The next week, the woman came to class with the entire family. She had followed my suggestion and challenged her family to a "longest Sit-Stay competition." First place won "a chore-free week with channel-choosing privileges," and last place had to take up the slack. She won hands down; her husband lost by a mile. He wanted revenge.

Oh! Baby steps! Don't you just love them? To me, training is all so predictable, but so many people fail to see or value the baby-step improvements early on in training and so give up. Quantification is a must to offer proof of improvement.

Testing for Comprehension

Training is all about effecting behavior change. Behavior is observable and therefore easily quantifiable. As training proceeds, it is essential to frequently test a dog's comprehension of verbal commands by checking response reliability. Rather than assuming that dogs understand our instructions, we should test that they know. And often, when we do quantify response reliability, we discover that dogs understand much less than we thought.

Dogs learn very differently from us. They are extremely fine discriminators and learn exactly what we teach; this is something we discover quickly when teaching cued behaviors. Unlike us, dogs have great difficulty generalizing what they learn to other situations. For example, a dog that learns a cued behavior taught by an instructor in class might not reliably perform that same behavior with the owner on the sidewalk.

All family members must practice in a wide variety of places and circumstances before dogs contextualize these requests to every possible scenario. Though this may sound daunting and time-consuming, it's not. The two best ways to get dogs to generalize instructions to all scenarios is to frequently interrupt walks every twenty-five yards, and dog play in dog parks every minute or so, with very short training interludes, most as short as a two- or three-second Sit-Stay.

I have always considered on-leash walks and off-leash dog parks and doggy daycares to be the most underutilized dog activities and facilities. They all offer perfect venues to teach high levels of reliability within enormous distraction, with the distractions ever-changing. This allows us to use the "distraction" of play as a reward — by saying "Go play" after each Sit-Stay.

Each time we test for comprehension, we find "holes" in training, and this provides our agenda for future training: fixing those holes. This way, bit by bit, we build an extremely reliable dog. Just as socialization and classical conditioning is for life and never ends, training and testing cued behaviors is ongoing. We must do it in numerous different scenarios with progressively increasing distractions.

We can never have a dog that is too trustworthy or too reliable because — news flash! — no dog or person is ever 100-percent reliable. We try our best to teach, and dogs do their best to learn, and if we keep teaching and testing, we have proof of improvement, whereas if we discontinue training, dogs will regress. Education is ongoing.

Response-Reliability Percentages

The nitty-gritty of what response-reliability percentages measure is this: Did the dog do it, and after how many instructions? Testing response reliability provides a revealing, objective index of performance that reflects both comprehension of commands and motivation to respond.

Lure-reward training is the only reward-based training technique where routine reliability testing is central to the process. Since lure-reward training was designed specifically to teach animals to respond *on cue*, we cue each behavior, and so the number of cues given and the reliability of responses are easy to calculate.

During the process of phasing out food lures and reliance on food rewards, make a habit of regularly testing off-leash and on-leash reliability of each of the six position changes in the S-D-S-St-D-St test sequence as well as using randomized sequences. Counterintuitively, when dogs are lure-reward trained, reliability is often higher when dogs are *off*-leash, compared to when on-leash. I think people's brains must wander when they put their dogs on-leash.

And here's a tip: When testing percentages, video your dog, whether you're at home, in the backyard, on a walk, or at a dog park. That way, you can focus solely on the test and cuing your dog, and you can review the video to calculate the numbers later, once you're home.

Just be prepared: When you first calculate response-reliability percentages, the results are often much lower than you would expect or hope. You may struggle to get scores that are even close to 50 percent. Even when testing dogs in the best possible training scenario — while you stand toe-to-toe with your dog in the kitchen with no distractions — most owners are amazed at their dog's poor responses to verbal cues. Then they find that response-reliability percentages drop by 20 to 25 percent simply by going outdoors, and then there's a further massive drop when other dogs are present, and then response-reliability percentages drop off a cliff when at a distance.

One finding that utterly amazed me from routinely testing dogs: When dogs have been trained on-leash, for many of them their proximal response-reliability percentages drop to almost zero when just six feet away from their owner *and on-leash*. Wow! Blew my mind. Next to zero control within the length of the leash. To me, that would seem to indicate that people who customarily walk their dogs on-leash don't train as much, whereas practicing recalls off-leash in puppy classes can considerably improve proximal reliability, whether off-leash or on-leash.

When I first tested ZouZou's Distance Sits from approximately twenty yards away in a dog park, her score was a little over 6 percent. This was very embarrassing considering I had an audience. But that's why we test. That score made clear that it was much too soon for us to let our newly adopted, fence-jumping, deer-chasing Beauceron off-leash in an unfenced area.

But let's look on the bright side: If your dog gets an extremely low initial score, they will certainly be in line to win the award for *most-improved dog*, just like our Jack Russell terrier with the 0.2-second Sit-Stay!

Testing Using Handsignals Only

To begin, take a step back and call your dog, and then signal the S-D-S-St-D-St sequence three times in a row. When testing for handsignals only, if your dog does not respond to the initial signal, emphatically *repeat the signal.* If after five handsignals, your dog has not yet responded, repeat using a food lure until the dog responds. However, each time you use a food lure counts as an additional cue.

After finishing each sequence, count how many handsignals (plus food lures) you gave for each of the six position changes. Then total the number of cues for the three repetitions for each position change. Here is an example from a very typical test.

S	D	S	St	D	St	TOTAL
1	3	2	5	2	8	
1	1	2	3	2	4	
1	2	1	4	3	5	
3	**6**	**5**	**12**	**7**	**17**	**50**

For each of the six different responses, divide the *number of repetitions* (three) by the *total number of cues* (in this case, all handsignals, apart from three lure-movements for the first Stand from Down to get the dog into position), and then multiply the result by one hundred. This gives you the response-reliability percentage (RR%) for each position change.

$$\textbf{RR\%} = \frac{\textbf{\# of Responses}}{\textbf{\# of Handsignals (+Lures)}} \textbf{ x 100}$$

	S	D	S	St	D	St	TOTAL
Fraction	3/3	3/6	3/5	3/12	3/7	3/17	**18/50**
Percentage	100%	50%	60%	25%	43%	18%	**36%**

The calculations are simple and straightforward. However, this simple index will accelerate training no end because, as you can see, it highlights both strengths and weaknesses. And each time you test again, you can compare percentages and monitor your progress.

In this example, Sit from Stand is perfect, and even early in training, that wouldn't be unusual. However, Sit from Down, Down from Sit, and Down from Stand need practice, and clearly, Stand from Sit and Stand from Down need lots of practice.

What I would recommend is, immediately after testing, use a food lure for half a dozen repetitions of the entire sequence. Then conduct the three-sequence test again with handsignals only, while also preferentially rewarding your dog for the last three position changes (both Stands and Down from Stand).

Testing Using Verbal Requests Only

When testing response-reliability percentages for verbal requests only, if your dog does not respond to the verbal request, calmly, clearly, and emphatically *repeat the verbal request*. If your dog doesn't respond after *five* verbal requests, follow each additional verbal request by a handsignal, and repeat until your dog complies. However, each handsignal counts as an additional cue. The equation for the calculation is the same as when testing handsignals only:

$$\textbf{RR\%} = \frac{\textbf{\# of Responses}}{\textbf{\# of Requests (+Handsignals)}} \times 100$$

I always recommend that people test their dog before and after each training session. That way, they can evaluate improvement in reliability from each session and graph end-of-day results to see trends. Dogs and people occasionally have off days, and you don't want a single bad performance to cause undue upset. If the overall trend is upward, then don't worry about occasional dips in scores.

Repeat the test in different rooms of the house, in your yard, during on-leash walks, and in dog parks. Having proof positive of your dog's degree of compliance in all these different scenarios provides you with a realistic view of your level of control over your dog in each situation.

Response:Reward Ratios

Training takes a quantum leap when you start to pay attention to the number of food rewards you use in each session. The *response:reward ratio* is a simple way to keep track of how many responses you are getting per food reward.

As a guideline, remember the average golden retriever puppy will gladly perform twenty consecutive food-lured Puppy Push-Ups for the prospect of a single piece of kibble. Since each push-up involves two responses, the response:reward ratio is 40:1.

I calculate response:reward ratios as soon as I start to train any puppy or dog. And as much as possible, I go through the same initial eight trials with every dog. I give a verbal request, and then lure each response with kibble. This sequence is not set in stone; you may proceed anyway you like. The following is simply what I've been doing with each new puppy or dog for years.

1. "Hello, Rover. I'm your trainer."	Then praise +1 reward
2. "Rover, Come."	Then praise +1 reward
3. "Rover, Come, Sit."	Then praise +1 reward
4. "Rover, Come, Sit, Down, Sit, Stand, Down, Stand."	Then praise +1 reward
5. "Rover, Come, Sit, Down, Sit, Stand, Down, Stand."	Then praise +1 reward
6. "Rover, Come, Sit, Down, Sit, Stand, Down, Stand."	Then praise +1 reward
7. "Rover, Come, Sit, Down, Sit, Stand, Down, Stand."	Then praise +1 reward
8. "Rover, Come, Sit, Down, Sit, Stand, Down, Stand."	Then praise +1 reward

Response: Reward Ratio for Trials 1–3: 3 responses for 3 rewards, or 1:1

Response: Reward Ratio for Trials 4–8: 35 responses for 5 rewards, or 7:1

Response: Reward Ratio for Trials 1–8: 38 responses for 8 rewards, or 4.8:1

Trials 1–3

I like to first introduce myself and offer a piece of kibble to check whether the dog is confident with me as the trainer. Approaching, sitting close, and accepting kibble from a stranger are three extremely powerful temperament tests. If the dog happily does all three, we're cool, and we're off and running.

However, if the dog doesn't readily accept the kibble, doesn't approach, and doesn't stick around, I'll ask the owner to offer a piece of kibble. If the dog also refuses food from the owner, I assume the dog is stressed by the scenario. However, if the dog readily approaches, sits, and takes food from the owner, after not doing so with me, then the problem is likely that the dog is not comfortable with me, which is a considerably more urgent and important issue. I'll delay testing position changes while I try to build the dog's confidence in the new setting and with me by offering treat after treat as I keep moving away from the dog. As I've mentioned before, lack of confidence can undermine any training program at any time.

Trials 4–8

Once we're up and running, I like to take a shot at luring the entire test sequence for a single food reward. I keep the dog's interest alive by frequently waggling the food lure and making it come alive. With puppies, my success rate is about 60 percent.

With puppies I'm not successful with and dogs, I take the time to practice luring some of the more difficult position changes, such as the Stands, Down from Stand, and Sit from Down. Then I retest.

Testing Both Simultaneously

For each test of response-reliability percentages for either verbal requests or handsignals, I count out three food rewards and repeat the test sequence three times in a row, offering a single food reward per sequence. Because I always count out the number of food rewards beforehand, I am always working with a response:reward ratio of 7:1, which means it will be very easy to further increase the ratio (and decrease the number of food rewards) to increase the reinforcing power of each food reward. After testing, I review the video to count the number of cues and calculate the response-reliability percentages for each position change.

Again, training and testing go hand in hand. As I test for response-reliability percentages, I am also training the dog. During each sequence, the single reward doesn't necessarily come at the end of the sequence. I usually preferentially reward the dog for one of the more difficult position changes (those with the lowest response-reliability percentages), or for an exceptional one.

As soon as you start quantifying your training, your dog's reliability will improve by leaps and bounds, and your dog's dependence on food rewards will become a nonissue.

Part 3

LURE-REWARD TRAINING STAGE TWO: MOTIVATION

The second stage of lure-reward training is about motivation, motivation, and motivation. The goal is to inspire dogs to *want* to do what we would like them to do to such a degree that dogs become self-motivated. Their responses become internally reinforced until "just doing it" becomes the reward, and dogs enjoy doing what we ask as much as they enjoy all their normal, natural doggy behaviors.

The first item on the motivational agenda is to increase the reinforcing power of food rewards by rewarding less frequently, unpredictably, and seemingly at random, which substantially increases the dog's anticipation. At the same time, we preferentially reward the dog for better-quality behaviors.

Then by integrating numerous very short training interludes into dogs' favorite activities, such as walks, sniffing, and dog-dog play, we use these activities as extremely powerful *life rewards* that mostly replace food rewards. Rather than letting these activities become distractions that work against training, we integrate training into them, so that the distinction between play and training becomes blurry, and training becomes just plain fun.

Finally, we also put "problem" behaviors on cue, so we can allow dogs to be dogs while managing the behaviors we find annoying — like jumping up, barking, and hyperactivity.

Chapter 8

Using More-Effective Reinforcement

The essence of successful lure-reward training is knowing *how often* and *when* to reward, so that desired behaviors are reinforced most effectively. I know it sounds counterintuitive, but *asking more for less*, or requiring more and more responses for fewer and fewer rewards, increases motivation. When dogs are given lots of food rewards, they are quickly taken for granted and lose their reinforcing power. For a reward to be maximally effective and have the greatest reinforcing effect, it must be *immediate*, appear to be *unpredictable* to the dog, and preferentially reinforce only *higher-quality* responses.

Rewards only reinforce behaviors that occurred within the last couple of seconds. Any delay drastically reduces the reward's reinforcing power. Of course, all rewards teach dogs to enjoy training and love their trainer.

Immediacy of reinforcement is most important when teaching speedy position changes, but when working with long-duration behaviors — like extended stays, Off, Shush, and lengthy recalls and heeling sequences — timing is not nearly as critical. Instead, praise and occasionally reward intermittently for the duration of the behavior, and especially right *before* the end of a lengthy behavior.

Obviously, avoid rewarding a dog immediately after a release command, which would send the message that you, too, are happy that the exercise has ended. When I release a dog, and especially when a training session is over, I walk away sad and gloomy. I want my dogs to feel the same, so they look forward to the next session.

Everyday Preconceptions about Reinforcement

When I was still giving regular seminars, I would ask lots of questions to gain insight into dog trainers' views regarding reinforcement. Their answers were fascinating and ranged all over the place. What I found was that most trainers rewarded too much, while others barely enough.

Here is the question I liked to ask most, which I rephrased in a variety of ways: Let's say you're going to practice ten recalls in a row to speed up a dog's recalls. To reinforce most effectively, after *which* of the ten recalls would you plan to reward your dog and why?

The answers varied considerably, but the most common fell into five categories: "all of them," "the first," "the last," "the best," and "half of them." Before reading my take on these approaches below, answer this question for yourself. What do you think would be the most effective approach to reinforce the speed of recalls?

- **"All of them":** Trainers often chose this "to keep the dog motivated." However, *continuous reinforcement* is not particularly motivating because dogs tend to take the rewards for granted, and so they lose their reinforcing value. Dogs are less motivated when rewarded for every response.
- **"The first":** Some said this "to get the dog off to a good start." However, what if a dog takes a while to warm up, and the first recall turns out to be the worst of the lot?
- **"The last":** Trainers who chose this wanted "to end training on a high note." Again, what if the last recall turns out to be slowest?
- **"The best":** This is a wonderful idea, but how do we know which is the fastest until the dog has completed all ten?
- **"Half of them":** To me, rewarding a dog for 50 percent of

> responses is a rich schedule — too many rewards. My ideal during early training is to reward more like 15 to 30 percent of behaviors. However, whether it's 15, 30, or 50 percent, rewarding "some" of a dog's responses reflects *intermittent reinforcement*. This is more motivating than continuous reinforcement, but *which* five responses should we reward?

There are several types of intermittent reinforcement. If we reward every other response, that would be a *fixed schedule* — and like continuous reinforcement, the rewards are predictable and therefore less motivating.

If we randomly choose the recalls ahead of time (say the second, fifth, sixth, eighth, and ninth recalls), that would be a *variable schedule* — which is unpredictable and very effective, so long as not too many rewards are given overall.

If we decide to reward only the fastest recalls, that would be a *differential schedule* — and it's a brilliant idea. But again, how do we know which responses are the fastest until the dog has completed all ten?

Amazingly, when I asked this question at seminars, one or two people always answered zero! I presume they misunderstood the question, which asked the best way to *reinforce* the *speed* of recalls.

Reducing Food Rewards

Without a doubt, the biggest problem that people run into when using any reward-based training technique is rewarding dogs too often. Offering food rewards willy-nilly not only reduces the value of food rewards in our dog's eyes but also in *our* eyes. Food rewards become more like rote participation ribbons or stickers, given every time just for showing up.

Now, I am not against rewarding participation, nor do I have anything against stickers and ribbons. People love them. Children love stickers and ribbons at school. I love them. Only yesterday, I put up three stickers of Samuel Squirrel in prominent places in my house.

However, any reward given routinely quickly loses its motivational value.

Back in the early days of puppy classes, I used to give performance ribbons to dog owners for winning games. I gave out only one or two ribbons a week, and sometimes none, but by the end of each six-week class, I made certain that everybody received a ribbon at some point. In a sense, this was a participation ribbon, but its presentation appeared to be based on quality of performance. The ribbons reinforced every person's participation in class because they seemed to be given for more than just showing up.

I also did my best to reinforce their presence in class in other ways, such as by making a point to frequently compliment their dog and offering well-earned praise for a job well done. Training shouldn't be "positive" just for the dog; owners need motivation, too. Owners are teachers in training.

Asking for More and Longer for Less

From the very outset of training, always ask for more and more responses and for longer and longer durations for fewer and fewer rewards. As you train, always keep track of response:reward ratios and progressively increase the ratio by starting with baby steps. After rewarding a dog for one Come-Sit for a single reward (2:1), try two Come-Sits for one reward (4:1), and then three, five, eight, and fifteen Come-Sits for a single reward (30:1).

Do the same thing as you practice S-D-S-St-D-St sequences: Start with just one reward per sequence, and then one reward for two sequences, then three sequences, and so on. It's all too easy and very, very effective. As you start to thin out food rewards, they become less predictable and hence, more powerful, and so dogs become even more motivated to do even more to get them. Similarly, the longer you keep a food reward in your pocket, or in your hand, the longer the stays, Shush, Off, and Heel.

Preferentially Reward Quality

As you reduce the overall number of food rewards, it becomes possible to reward only the dog's *better-quality responses*. View a food reward as an Olympic medal, and you are the judge. Based on your dog's performance, did they qualify for the Olympics? Did they get on the podium? If so, which medal did they win?

First consider whether a response is even reward-worthy. There's not much point in rewarding sub-par responses. Praise for responses that meet specific minimal criteria, offer more and/or better rewards for better responses, and even more and/or the very best rewards for the best responses. Try to make rewards mirror the quality of your dog's responses.

Move away from the same-old-single-piece-of-kibble food reward for every response. Don't be like a vending machine but more like a slot machine, so your dog never knows for certain when a reward is coming or what it might be. Better yet, be more like a video game that specifically rewards players according to the quality of their play.

Be Magical

Dogs always know from a distance whether we have food in our hand or pocket. When an owner wears a belt with a bulging bait bag, two tennis ball clips, three tug toys hanging down like gopher pelts, and a Frisbee stuck in their shirt, it is obvious that fun times are nigh, and the dog gladly engages. However, is your dog similarly attentive and responsive when they know you have no food and no toys?

To maximize reinforcement, we want dogs to learn two essential things:

- First, just because we have food in our hand or on our person doesn't necessarily mean they will get it — not immediately, and maybe not at all.
- Second, just because we *don't* have food in our hand or in our pocket doesn't necessarily mean that dogs won't occasionally receive a food reward (for an exceptional response).

It is a lightbulb moment for dogs when, even though we don't have food on our person, we make a food reward appear as if by magic.

For instance, I used to station a few food rewards around the house — in cupboards, cabinets, and drawers, on tables and the mantelpiece — and then I'd wash my hands and let Phoenix sniff to affirm no food was present. Then we'd practice recalls, heeling, and stays. Occasionally, I would grab a previously hidden food reward seemingly out of thin air, like a white dove from a top hat. Magically plucking treats out of the air impressed Phoenix no end, and so I also hid some treats in the garden.

One day before our walk to Codornices Park, I hid treats along our intended route. Phoenix was amazed when, after a quick Come-Sit, a vibrant heeling sequence, or a lengthy Down-Stay, she was given food rewards that I reclaimed from a wall, a packet pinned to a notice board, and from under a rock. Each time she took a piece of freeze-dried liver, she sniffed my jacket to confirm that I had no food on me. But there were more doves in the hat.

It was the tree that left Phoenix stunned in wonderment. I must have plucked a half dozen bits of liver from that tree to periodically reward her for her longest Sit-Stay on record. This single experience left an indelible impression on her Malamutian mind because, on many walks thereafter, she would always pause and gaze up at the treat tree with her nostrils flaring, nose-scanning for more treats. I used to tell her, "Wrong season, Phoenie. We'll have to wait till spring."

Through those chill, short days of winter, I practiced the art of catching indoor "air treats." They flourished in the drawing room. "Phoenie, Look! There's one." I would stand up and walk round the room and point, and Phoenix would follow, and then I would make a grab: "Got it! Mmm-mmm-mmmmmm! This one's gooood, Phoenie." I pretended to nibble at the imaginary treat before saying, "Fancy a taste? Phoenix, Sit." When she did, I would offer her the remains of the air treat — and she would actually eat it! Used to crack me up every time.

My "air treat" record was at a Halloween party where a cute pit bull cross ate twenty-two air treats in succession in front of the costumed guests. It was a carnival moment. Try this yourself.

Use all these techniques to enhance motivation: Reduce the number of food rewards, reward only deserving behaviors, vary the amount and type of rewards, and make food rewards magically unpredictable — so dogs will never take them for granted.

Food Rewards Have Lifetime Uses

Even though we need to reduce the number of food rewards when teaching basic manners, they have many lifelong uses. For example, socialization, handling, and desensitizing subliminal bite triggers and scary environmental stimuli never ends; a single classical-conditioning session may require several hundred food rewards.

Moreover, food rewards are the easiest way to transfer *your* training and control to others. An enormous asset of lure-reward training is that it is easily transferred to other family members (especially children), visitors, and strangers.

And don't forget that dogs love food rewards, and we love giving them. So long as food rewards don't undermine training, it's perfectly fine to give them to your dog, even as life-participation awards — for just being there. While writing this book, I frequently needed to pause and think, and at the same time, I gave each dog a couple of pieces of kibble for lying down quietly next to me and providing the necessary inspiration.

Also, I often give a few pieces of kibble when dogs complete a regular daily task, such as the final de-pee and de-poop safari of the evening. I offer praise and offer a few pieces of kibble outside as soon as they produce, varying the number according to speed and production, and then back indoors, the dogs all join the treat circle near the Ziwi jar on the counter and each receives a piece in turn (three rounds), before walking to their respective beds, whereupon once settled, they each receive three good night kibbles.

Types of Reinforcement Schedules

Basically, a reinforcement schedule dictates *how many* responses and *which* responses to reinforce by offering a reward. Whether or not to reward an animal following desired responses has been the subject of thousands of scientific studies, and these studies have inspired a number of reinforcement schedules. Most reinforcement schedules were developed and refined in research laboratories, and most studies involved computers training caged rats to maximize work ethic. Most studies used a single food pellet as a reward.

The various schedules for using a single food reward over and over were later applied to most aspects of animal training. Additionally, the principles behind these schedules have been universally entrenched in several human fields, including child raising, education, motivating employees, and so on.

Basically, there are *continuous* reinforcement schedules and several *intermittent* schedules, which can be either *fixed*, *variable*, *random*, or *differential*. Below, I evaluate these schedules in terms of their relative effectiveness for motivating dogs (and people) as well as their ease of application.

Continuous Reinforcement

Continuous reinforcement means *rewarding the dog after every response in a long sequence*. For example, giving a reward after every sit or every recall.

Understandably, during early training, when dogs are rewarded with treats often, they quickly think they are in treat heaven. However, excessive food rewards quickly reduce a dog's perception of their value, and after the flush of initial success, trainers find that dogs begin to take food rewards for granted. Moreover, since food rewards are predictable and expected, there is no immediate urgency for the dog to respond, and so promptness and response reliability falter. In some scenarios, slow recalls become no recalls.

Many people think that the more rewards they give, the better the motivation and performance, but the opposite is true. Of course, dogs thoroughly enjoy a smorgasbord of treats, and love us for it, but they won't learn much. It is so important to maintain and progressively increase the reinforcing power of food rewards.

Also, continuous reinforcement makes it agonizingly difficult to reduce or phase out food rewards. Some dogs go on strike the first time a response is not rewarded. Think about it: If a food vending machine didn't deliver when you put in a dollar, would you put in another dollar? Precisely. And how happy would you be with a vending machine that didn't deliver your morning cup of Joe? Surely, we don't want our dogs to feel the same about us.

From a behavior-modification perspective, continuous reinforcement is hopeless.

Fixed Reinforcement

Fixed reinforcement means *rewarding the dog after a set number of responses (fixed ratio) or a set amount of time (fixed interval) in a long sequence*. For example:

- With a *fixed ratio schedule*, a dog might be rewarded after every third recall.
- With a *fixed interval schedule*, a dog might be rewarded after every five seconds of silence following a request to Shush.

The big drawback of fixed schedules is that, like a continuous schedule, they quickly become predictable. Thus, when used for any length of time, they lose motivating power and undermine reinforcement.

That said, I use a fixed schedule initially with baby steps, for example, when trying to increase response:reward ratios when teaching body-position changes or lengthening attention and stays. The key to success is to use a fixed schedule for only a few repetitions and then stretch the schedule. For example, I might ask for one Heel-Sit for one piece of kibble and repeat this six times. Then I'll ask for two Heel-Sits for one kibble six times, and then three Heel-Sits, and so on. In essence, I am gradually upping the fixed schedule every half dozen repetitions.

However, trying to *maintain* healthy rates of responding for *lengthy periods* with the *same* fixed schedule doesn't work that well. It isn't just that rewards become predictable and less reinforcing. When rewards are tied to productivity, so that the dog must perform a certain number of responses before getting "paid," the dog tends to rush through responses in their eagerness to meet the quota, and the quality of responses takes a hit.

Fixed interval schedules allow dogs to predict *when* a reward will come, which causes fluctuations in work ethic and the quality of responses. Dogs focus more and pay increasing attention as the time for an expected reward approaches, but immediately following the reward, the dog's attention and quality of response take a big tumble.

Moreover, if a fixed schedule becomes *well-established*, then changing the schedule to ask "more for less" can be disastrous. The dog might *go on strike or resign* and simply refuse to do what's asked.

Again, consider the human equivalent: If our paycheck suddenly came less often, or our salary were reduced, would we work just as hard or even stick with our job? Indeed, given the lack of effectiveness, unreliable performance, and all the problems associated with long-term fixed schedules, I find it truly surprising that *the entire workforce of the world has been maintained for years and years on fixed schedules — piece rate and payday!*

Paying employees on a fixed schedule is an obvious legal requirement. However, motivating employees is a different kettle of fish. How about a treat-tree in the foyer with occasional surprise presents? Or an occasional box of chocolates in the receptionist's filing cabinet? Perhaps one day the boss could come to work early to serve coffee and Danish to every employee as they arrive to work. Or the office could close early one Friday afternoon during the holidays so that employees could shop and buy presents for family. Or what about a bone, flirt pole kitty toy, or a good book for employees who have dogs, cats, or children? It's the little thoughtful things that make the workplace a productive village. Things as simple as a single word of acknowledgment and appreciation.

Variable Reinforcement

Variable reinforcement means *rewarding a dog on average after a set number of responses (ratio) or on average after a set amount of time (interval), with the number of responses or durations varying between rewards.* For example:

- With a *variable ratio reinforcement* schedule, a dog might be rewarded *on average* after every three Come-Sits during a sequence of ten, but the number of Come-Sits between each reward would vary.
- With a *variable interval reinforcement* schedule, a dog might be rewarded *on average* every five seconds during a thirty-second Sit-Stay, but the number of seconds between each reward would vary.

Each reward is completely unpredictable, and therefore unexpected, and this adds a lot of spice to the training game and captures a dog's attention. As dogs anticipate the next surprise reward, their brains get bathed with feel-good dopamine with an occasional surge of adrenaline. Basically, through anticipation, dogs become addicted to the training game.

Variable schedules make it much easier to ask for more and longer for less. Variable schedules intersperse a series of unreinforced responses with occasional reinforced responses, but with no way for the dog to predict when rewards will arrive. Thus, even when dogs are asked to keep working through progressively longer strings of unreinforced responses, they don't become discouraged or go on strike because they've already learned that rewards might come at any time.

As an added benefit, since rewards are unpredictable, dogs don't rush consecutive responses. Instead, they focus better and work steadily, and responses become sharper and more stable. Variable reinforcement schedules are outrageously effective for maintaining a solid work ethic. Just think how long people will play slot machines and video games.

If this sounds too good to be true, you're right. There are two big downsides. One is, calculating variable schedules is difficult for a lot of people. And second, it's difficult to train a dog in real time while calculating or even following one.

Calculating Variable Schedules

I love playing with numbers. I love calculating autoshaping algorithms or computing variable and differential schedules. However, most people don't, and for good reason. It can be complicated, and most people are more interested in training their dog than doing math.

However, as an example, here is a variable schedule to reinforce position changes: eighteen position changes, three each of the six different position changes, and rewarding the dog for six of them (33 percent of responses); in other words, this schedule rewards the dog on *average* after three position changes (response:reward ratio = 3:1). The rewards appear to come at random, and so are unpredictable. The second line uses letters to indicate the sequence of positions; bold letters indicate the six position changes to be reinforced with a reward.

	1	2	3	4	5	6	7	8	9	10	11	12	13	14	15	16	17	18
Standing →	**D** -	St -	**D** -	S -	**St** -	S -	D -	S -	D -	**St** -	S -	St -	S -	D -	S -	**St** -	D -	**St**

As you can see, the eighteen position changes don't follow a set pattern, nor do the six being reinforced — the first, third, fifth, tenth, sixteenth, and eighteenth. Rewards appear at variable points throughout. Nevertheless, the order of the position changes and those that are followed by a reward are only *seemingly* randomized. Everything is arranged to preferentially reinforce two each of the three more difficult position changes: Stand from Down, Stand from Sit, and Down from Stand.

Yikes, I hate myself when I write this stuff, but I've done so to make a point: This is all so needlessly complicated for the average person.

Variable schedules were created and applied by laboratory computers, and their principles are effectively and profitably applied to autoshaping and to slot machines and video games. Oh, these schedules are effective alright, as we know from the addictive nature of 24/7 operant conditioning (electronic autoshaping), gambling, and gaming. However, people are not computers. Few people can calculate and apply variable schedules and train a dog at the same time. Booooo!

But there's good news: if we want rewards to be unpredictable and appear to come at random, *why don't we just reward at random?*

Random Reinforcement

As the name indicates, random reinforcement means *rewarding a dog at random.*

When weighing the highlights and downsides of all the reinforcement schedules so far, it's clear that *consistency* can quickly undermine reward-based training. Whereas consistency is absolutely essential when using aversive punishment, consistency is neither essential *nor advised* with reward-based training. Instead, since unpredictability is the key to anticipation, effective reinforcement, and motivation, we can utilize our natural human *forte — inconsistency!* Why don't we just intentionally try our best to reward our dogs at *random*? Consistency *not* required. Computing power *not* required.

I use random reinforcement with every new adult dog I train. That leaves my brain free to concentrate on assessing the dog's behavior and temperament and evaluating how quickly the dog "joins up" and pays attention, approaches, and stays close.

To avoid rewarding too frequently, think what you're planning to teach and for how long, and count out the rewards beforehand with the aim of rewarding your dog for roughly a third of responses, and then off you go: *Reward at random* until you run out of rewards.

Video each session for review to confirm response:reward ratios, calculate response-reliability percentages, the durations of stays, and so on. Challenge yourself to improve each session.

Random reinforcement is amazingly powerful; it combines *all* the advantages of variable reinforcement, since it creates unpredictability, and... *it's pretty easy.* For training to be successful, it *must* be easy, for both dogs *and* people.

If this too sounds too good to be true, you're right again. In fact, random reinforcement shares the same big downside as *all* the reinforcement schedules discussed thus far. None of these schedules consider the *quality* of responses.

In essence, with continuous, fixed, variable, and random schedules, we *reinforce just as many below-median quality responses as above-median quality responses*, which is plain silly. But that is a failing of how they were conceived. All were designed to increase motivation and work ethic, not to improve qualitative aspects of behavior. But in dog training, quality is also important. We want to improve speed, precision, pizzazz, and style. Consequently, we need qualitative reinforcement criteria.

Differential Reinforcement

Differential reinforcement means *rewarding a dog only if their response meets or surpasses a set criterion, offering better rewards for better responses, and reserving the very best rewards for the very best responses.*

A lot of the time, reward-worthy recalls are blatantly obvious. For example, when hiking a trail with your dog, who is twenty yards ahead and greeting another dog, you see people approaching on horseback, so you call your dog, and in one blurry flurry of activity, your dog terminates their butt sniff, explodes toward you at warp speed, butt-slides into the most beautiful Sit-Front you've ever seen, and arches their neck upward to look you in the eye, as if to say, *You called, Ma'am?* Now *that's* reward-worthy!

However, to go back to my original question: If we want to improve the speed of recalls, *which* of ten recalls would we plan to reward, *why*, and *how*? The inherent dilemma with differential reinforcement is that we never know for certain whether a response is passable, better, or best until the dog has completed all responses in an exercise.

Consequently, beforehand, we must test and calculate average responses. Then we can set precise criteria for which responses are reward-worthy, and if so, how reward-worthy. Then, when training, we can know in the moment if any response exceeds any of the three criteria for reward-worthy, better, or best responses.

Setting Criteria

As an example, let's establish a minimal criterion for reward-worthiness for speed of recalls — a Come-Sit from forty-nine feet. (This is the racing distance in the K9 Games that is fairest for most breeds; it's the distance a golden catches up with a border terrier.) First, we need to determine the *average* recall time over ten trials. We could do this in a single session or spread the ten recalls over the course of the day for quick-to-tire dogs. To calculate the average recall time, add up the recall times for all ten recalls and divide by ten.

But first, let's pause and ponder what we might expect: How many of the ten recalls would you *expect* to be *quicker* than *average*? Take a moment and write down a number from one to ten. This was another question I used to ask at my seminars.

From seminar audiences, the almost unanimous answer was five, which by definition, of course, would be the number of recalls faster than the *median* recall time. My guess is that, most likely, seven or maybe eight recalls would be faster than the *average* time.

This is the answer we should expect with an open-ended scale. For example, a dog cannot be much quicker than quick, but a dog can easily be a whole lot slower than slow. It only takes one or two extremely slow recalls to pull the average away from the median. As an example, here is data I adapted from testing basset recalls:

Times of ten consecutive recalls:

1	2	3	4	5	6	7	8	9	10
10.2	**9.1**	**8.5**	**9.6**	**9.9**	**7.5**	**9.4**	**17.6**	**6.8**	**11.4**

Median = 9.5 (50% of responses quicker than median time)

6.8	**7.5**	**8.5**	**9.1**	**9.4**	9.6	9.9	10.2	11.4	17.6

Average = 10 (70% of responses better than average)

6.8	**7.5**	**8.5**	**9.1**	**9.4**	**9.6**	**9.9**	10.2	11.4	17.6

I prefer to only reward a dog after 15 to 30 percent of responses on average. Rewarding after every above-median response would be a rich schedule (50 percent), and rewarding after every above-average response (70 percent) would be *even richer*, therefore losing more reinforcing power.

By "reverse engineering" the question from real data, I decided to make *10 percent better-than-average* the *minimal criterion for reward-worthiness*. In this example, the average recall-time is 10 seconds, and only three responses are quicker than 9 seconds (10 percent better than average), which is ideal. Now, since we're creating a differential reinforcement schedule, we also needed to establish *additional criteria for better and best responses*:

A standard reward for recalls:	quicker than 9.0 seconds (10 percent quicker than average)
Better rewards for recalls:	quicker than 8.0 seconds (20 percent quicker than average)
Best rewards for recalls:	quicker than 7.0 seconds (30 percent quicker than average)

Now we are ready to train. All we need to do is remember these three reward criteria, time every recall with a stopwatch, quickly check if the recall is quicker than 9.0, 8.0, or 7.0 seconds, and reward accordingly. I usually practice recalls in batches of ten, and after each batch, average the recall time, recalculate the three time-based reward criteria, and plot the results on a graph. Always keep records. Proof of improvement is so motivating.

Further, modulate your praise and the value of rewards to match whether they are good, better, or best. Typically, some of the better responses, and all the best responses, will be exceptional, so heartily celebrate with your dog.

Be Random but Discerning, and Reward the Ineffable

Differential reinforcement will seem random, since rewards are based on a dog's variable performance. However, for a lot of people, a differential reinforcement schedule has the same drawback as variable reinforcement: Although highly effective, it requires mathematics.

And so we arrive at *the easiest to apply, math-free, most-effective reinforcement schedule*: a combination of random and differential reinforcement.

Count out your food rewards beforehand, so that you do not over-reward, and then reward at random (or try to), but each time, right before you do, ask yourself two quick questions: Is this response *really* reward-worthy? And if so, *how* reward-worthy? Impose a quality override to random reinforcement, either based on preestablished criteria or your own intuition.

Even better, it's OK to change your mind. It's good to change your mind. Maybe you haven't rewarded for five responses in a row and feel it's time to do so, but the next response just isn't up to snuff, so don't. Or you just rewarded twice in a row, and you want to wait before rewarding again, but the next response is off the charts, better than you've ever seen. Go for it! Heap praise and reward like crazy.

Vacillation works for you, by magnifying your dog's anticipation and sending the reinforcing quotient through the roof. You grin, your hand dives into your food reward pocket, but only slowly, tentatively emerges, and as you dither over how much to reward, your dog's anticipation skyrockets as they await the judge's final decision: And the winner is... in the category of... *best moving... Distance Sit...* by an adolescent Bernese... Samuel... Squirrel!

The truly wonderful thing about random reinforcement, as long as we don't reward too much, is that we can just go with the flow and be ourselves, since discernment is one training skill where we humans excel. We may not be able to calculate as quickly or as effectively as computers, we'll never be as consistent as them or have their exquisite timing, *but*... we know what we feel. And we can quickly and easily *make snap judgments about the ineffable*.

Sometimes there's a "thing" about performances that cannot be objectively measured. For example, it's not so much that a dog looks at us for so many seconds, but *how* our dog looks at us. The other day, I saw a dog trotting down the sidewalk alongside their owner. *How* the old dog occasionally quickened their pace to keep up and *how* the dog looked at their owner made my day, as if the dog were saying, *Old bones. Trying my best. Love you.*

When training your dog, think of yourself as a judge on *America's Got Talent.* Mentally evaluate each response on a ten-point scale for *technical merit* and the *ineffable*, and then make a snap decision. Humans excel at instantaneously recognizing cuteness, style, panache — all those special, je ne sais quoi aspects of performance. In an eye blink, we can evaluate zillions of teeny pieces of sensory input and based on our lifetime of experience come up with an instant yes/no decision. Sometimes, we can't say why exactly, but we just respond, "Wow! Phenomenal! Stunning!" Trusting your intuitive judgment can make for very precise and powerful reinforcement, indeed.

Chapter 9

Using More-Powerful Rewards

The ultimate goal of training is to teach dogs to become so self-motivated that they *want* to comply whenever we make a request. While food rewards are a tremendous help during initial training, continue to be extremely motivating when used judiciously, and have invaluable uses throughout a dog's life, to achieve genuine self-motivation requires mostly replacing food rewards with much more powerful *life rewards*, such as walks, sniffs, play with other dogs, and playing interactive games with us, such as fetch, tug, and so on — all the things that dogs absolutely love doing. The secret is to integrate short training interludes into all these activities so that play and training become indistinguishable from each other. Training then becomes an effortless enjoyment — the best part of loving life with our dogs.

First, though, we must reinvigorate praise. Compared to computers training rats in cages (the source of theoretical learning principles), our best asset, when we teach dogs and other animals, is our voice. In fact, using our voice is the attribute that turns training into teaching. We need to use our voice more. Our words are instructive, but equally as important, praise is the quickest and *richest* means of communicating our feelings and gratitude to our dogs.

Additionally, we can take phrases like "good dog" and "thank you," which are potent primary reinforcers in their own right, and turn them into the most powerful *secondary* reinforcers on the planet.

Praise, Praise, Glorious Praise!

I often wonder what happened to our voice in dog training, not only teaching ESL to provide clear instructions from the outset but also for providing ongoing feedback: guidance, correction, and above all, gratitude. We simply must resuscitate *praise, praise, praise, and more praise!* Dog training without praise is like trying to grow plants without water.

Our tone of voice alone makes a huge difference. High-pitched, squeaky, baby-voice praise is intrinsically exciting for dogs. However, when praise is too exciting, dogs can become too excited and lose their self-control. Calm praise, in a normal heartfelt voice, is intrinsically soothing and calming for dogs. When teaching short-duration behaviors, such as body-position changes, I like to use a calm "goood dog" and an affectionate scratch behind the ear. When teaching longer-duration behaviors, such as stays, I like to give a running commentary that modulates according to the dog's attention, precision, and stability. Similarly, I give running commentaries when teaching dogs to follow and heel, and especially when dogs are playing together — ongoing feedback as to how well the dog is doing, plus lots of guidance to tone down the play style and prevent the energy from amping up.

For above-and-beyond behaviors, *praise from the heart.* Use multiple adjectives and linger over each word to mesmerize your dog: "Gooood lad!" "Wonnn-derful! Glorrr-ious! Extraaa-ooorrrr-dinary!" "Brilliant! Splendid! Sensational!" "Who's the smartest, slickest, styyylish girl?" "Whattta bright boy!" We should never be stingy on superlatives when praising our dog for getting it right!

However, there is one expression that I absolutely love to use and hear in any interaction, whether with animals or people: "Thank you." When said softly and meaningfully, it projects such enormous feelings of gratitude, recognition, and acknowledgment.

Of course, when delivering instructions, you must be clear and precise, but when praising, there is no need to hold back. Dogs don't understand much of what we say, but when they hear the affection and appreciation in our voices, they can feel the affection that's in our hearts, and so they, too, get an affect boost. Consequently, they wag-gaze at us with soft eyes, head cocked to one side, and maybe a flew-puff or two for good measure, which causes our hearts to melt as we feel their affection.

I just love thanking my dogs sincerely to express my gratitude and talking to them to discuss their behavior. I never get embarrassed. Of course, some people are a little more reticent to talk to their dogs and express emotion, especially in public. But we don't have to squeal and squeak when praising our dogs; we just need to be sincere.

Walking, Sniffing, and Play

Walking, sniffing, and dog-dog play are the biggest life rewards in domestic dogdom. If dogs had their druthers, they would likely walk and sniff forever and play until they drop. One of the joys of living with dogs is watching them indulge their noses and play with other dogs. However, during adolescence, sniffing and dog-dog play start to *compete with us for our dog's attention*. Usually, sniffing and play win by a landslide and quickly commandeer our dog's brain. Sniffing progressively invades walks and squeezes out forward progress, and at the dog park, we watch as our training progressively evaporates before our eyes. Our dog becomes slow to respond to requests, and surprisingly quickly, slow recalls and slow sits become no recalls and no sits. At this point, sniffing and play effectively reinforce the dog for *not* paying attention, *not* coming when called, and *not* sitting.

Moreover, it only takes one or two dog park visits for a dog to learn that the request "Rover, come here" signals the end of their outrageously enjoyable, social-and-olfactory extravaganza. When arriving at a dog park, after letting their dogs off-leash, most people chat or check email, and then once they are done, they call their dog. The dog comes running, happy that at last their person is ready to participate in the party! But the owner snaps on the leash and frog-marches the dog home.

Now, if dog-dog romping and a chance to sniff are the most joyful experiences ever, then terminating said good times must surely be the most devastating disappointment of all time. It doesn't take a Zen master to work this out, so the dog learns: Best to stay away from your person and dodge the collar grab or it's curtains for fun-time, especially if they say those dreaded words *Rover* and *come here.*

And what do people do to rectify the dog's lack of attention and compliance? They scream, "*Come here, you miserable cur!!*" And when the dog eventually comes, the person is angry, in word and manner, if not with physical correction. How else can the dog interpret the owner's actions, other than, *They're always angry when I go back to them. Always* praise your dog after a recall, no matter how long it takes.

As frustrating as this may be, the solution is surprisingly simple. If you can't beat 'em, join 'em.

Integrate Training into Life Rewards

Use the desire to sniff and play to your advantage by using these activities as life rewards to reinforce training rather than undermine it. To harness the reinforcing power of life rewards, integrate oodles of very short training interludes into all your dog's favorite activities, especially on-leash walks, sniffing, and off-leash dog-dog play.

For the most part, a training interlude may be as short as a single, short Sit-Stay-Watch, just to check that you have your dog's attention, then immediately say, "Let's go," "Go sniff," or "Go play." Sometimes, have longer training interludes to practice a few randomized position changes with variable-length stays in each position, or a few short Come-Sit and Heel-Sit sequences.

Frequent interruptions teach the dog that sitting and coming when called are not the end of the world, nor the end of the walk or dog park visit. Rather, they are short time-outs for a few kind words, some massage for the canine athlete, and maybe a refreshment break before it's back to walking, sniffing, or playing again.

Most important, the more times you interrupt your dog, the more times you can say the words your dog most wants to hear: "Let's go," "Go sniff," and "Go play." Moreover, resumption of your dog's favorite activities may be used over and over and over on any reinforcement schedule you like, and nothing will crack your dog's motivation, enthusiasm, and desire for more. Your dog will never tire of these rewards. Life rewards are life itself. The only difference is that walks, sniffing, and play are no longer competing against training. They're now reinforcing training.

On-Leash Walks

Walking in a straight line, at the same speed, and along the same route often bores dogs, and being bored, they quest sensory input, and we gradually lose their attention: *Yea! Odors are different... over here!* Zoom, the dog is off to the bushes.

To keep your dog's attention, change speeds frequently (going slow, medium, and fast), change direction frequently (usually restricted to about-turns on sidewalks), and most importantly, *stop at least every twenty-five yards and ask your dog to Sit-Watch.* Then immediately say, "Let's go," and resume walking. After twenty or so sits, you'll find that your dog looks up when you slow down to check whether you're going to stop. Eventually, your dog will automatically sit by your side whenever you stop. Praise, "Gooood dog," and then say, "Let's go!" and the walk resumes once more.

What if your dog doesn't sit? Then don't move. Switch to wait-and-reward training. Just wait until your dog sits and makes eye contact, and then say, "Let's go." It's that simple.

If you follow this routine on a three-mile walk, that would be 211 Sits, and each sit will be effortlessly and enormously reinforced by *resuming the walk.* Moreover, each sit occurs in a different scenario: quiet leafy streets, suburban streets, bustling downtown, next to schools and playgrounds, or next to a dog behind a fence. This practice is the quickest and easiest means to get a dog to generalize that Sit taught indoors is the same request in any and all situations. Walk back and forth *outside* a dog park fence for ten minutes before continuing the *on-leash* walk inside the dog park, and then integrate numerous short Sits into off-leash play.

Sniffing

Dogs love sniffing, and what better place than on a walk? However, there's no need to let your dog's urge to sniff bug you. Simply identify your dog's top ten favorite sniff spots and use them to reinforce training.

When approaching a favorite spot, ask for Sits more frequently, every couple of yards. When close, ask your dog to Sit-Stay and savor the moment. Let the myriad smells increase your dog's anticipation. Then say your dog's name, and the instant your dog glances at you, say, "Go sniff." What a reward! After four or five sniff spots, your dog will become well aware that *you* are suggesting and allowing the opportunity to sniff.

On my favorite walk, my dogs have three default sniff spots. *My* favorite is a lovely open area that is explored and anointed by many dogs. I let my dogs indulge themselves for ten minutes or so, sometimes much more, while I sit on a bench overlooking a canyon and just think and daydream. I have never considered a walk as something to hurry. I like to savor the moment, too. On lazy days, I'll dawdle and let the dogs sniff a good thirty or forty times.

On walks, you'll likely meet other dogs. Sniffing another dog's rear end is definitely on the top-ten list of all-time most-powerful canine life rewards. Also make this a gift from you. Ask your dog to Sit-Stay, check with the other owner that it's OK for the dogs to greet, and if so, instruct your dog, "Say hello." This is my polite, euphemistic way of saying, "Sniff butt."

Dog-Dog Play

For most dogs, dog-dog play is the be-all and end-all of life rewards. Running, chasing, being chased, and wrestling are the very essence of being a dog. Don't let this magnificent training opportunity go to waste; otherwise, play will likely become the *biggest* distraction to training. Regularly interrupt dog-dog play, ask for a short Sit-Stay-Watch, and then cue your dog to resume play. This is the quickest and best way to reinforce off-leash reliability, which radically changes your dog's quality of life for the better.

A dog park is one of the best places to use all the available life rewards at once. The more times you interrupt dog park play, the more opportunities you have to reinforce training by saying, "Let's go," "Go sniff," "Go play," or "Say hello." Every time you interrupt dog park shenanigans with a quick Come-Sit or a Distance Sit (see chapter 13), the *resumption* of dog park activities reinforces your brief training interludes. Get in the habit of interrupting your dog's play every couple of minutes. Additionally, I like to take a few longer breaks for the two of us to walk around *together*, or to settle down on the grass together and watch the other dogs playing. ZouZou would love it when I would make silly comments about the other dogs.

Dog Park Recalls

When I give this advice to people, I'm often met with the cry, "But *how*? My dog doesn't pay *any* attention to me when we're in the park." I can almost hear readers pulling their hair in frustration. If your dog has already learned that, in a dog park or any off-leash situation, "Come here" means "We're leaving," then, yes, your dog might not come to you until they are good and ready. But think about it. *All* dogs do return eventually. Otherwise, no one would be able to leave a dog park with their dog.

A vizsla owner once moaned about this during a behavior consultation: "My dog never comes when called in the dog park."

"Really?" I said and pointed to the vizsla. "Is that your dog?" The man agreed. "When did you last go to the park?"

"Just before I came here, so he would be calmer."

"So he came to you?"

"Well, yes. But he took forever."

"But then you put him on-leash to come here. There lies your problem."

"What? Coming here?"

"No. Putting him on-leash and leaving the park. Fenced dog parks are tailor-made for teaching reliability in an extremely distracting but safe setting. But sometimes you must wait until your dog is ready."

On park visits, wait and be ready for when your dog is ready. It's like, "When the pupil is ready, the teacher will come." In this case, "When the pupil comes, the teacher is ready." Regardless of how long it takes for a dog to come close enough for you to attach the leash, that's the time to start training. Don't immediately leave the park and go home. Attach the leash, leave the park, and then ... abruptly about-turn and walk back inside the park, ask for a Sit, offer five food rewards (one at a time), detach the leash, let go of the collar, and say, "Go play." Then walk or run away from your dog.

Most dogs will likely follow or give chase because, just a couple of minutes earlier, they came to you, and you were able to put on the leash and leave the park. But now you're back in the park, *and your dog is ready to train*. So when your dog comes, offer five more pieces of kibble and walk or run away. Repeat this half a dozen times, and then next time, call your dog, "Rover, Come." Most likely your dog will come when called, and so repeat recalls every half minute or so. You'll find you have a different dog.

Next park visit, do the same: (1) Say, "Rover, Sit," and lure the Sit if necessary; (2) take hold of the *collar* (to prove that you can); (3) *praise* enormously; (4) handfeed a few pieces of *kibble* one at a time; and (5) say, "Rover, Go play," and walk away.

On evening dog park visits, plan to handfeed your dog their dinner, course by course, in the park.

When your dog leaves your side, keep walking *away* from your dog. A receding owner is far more alluring than one sitting on a bench with their eyes glued to a phone. If your dog doesn't come back after wandering off, wait until they do. All dogs return eventually, so just keep walking and wait. Or, hide, watch, and wait.

When new dogs enter the park, use this opportunity to practice recalls from play. Stay close and when your dog has finished greeting or playing with the other dog, call your dog and walk away. Go through the Sit-and-kibble routine, and always remember to take hold of your dog's collar before rewarding.

By doing all this, you are no longer an observer but a major participant in your dog's dog-park experience! It is so important that *you and training* become intricately integrated into all your dog's favorite activities.

When it's time to leave for real, give your dog an end-of-play stuffed chewtoy, either back in the car or when you get back home.

Interactive Games

Playing interactive games with dogs — like fetch, tug, and hide-and-seek — is the fast track to off-leash reliability. For some dogs, playing games becomes a passion that may rival walks, sniffing, and dog-dog play.

All three games increase a dog's desire to approach and stay close, which greatly facilitates teaching off-leash following and heeling, and hence, loose-leash walking. Of course, during all three games, recalls appear by magic. Dogs learn extremely quickly that after catching a Frisbee or grabbing a bouncing ball, they need to return the object to their person for it to be defibrillated so that it can bounce or fly once more. Similarly, tug is more fun with a person holding the other end, and the whole point of hide-and-seek is for the dog to find you.

Interactive game toys become more alluring if reserved exclusively for dog-human play. After a game, put the toys away. If dogs get to keep interactive toys, this quickly devalues their specialness, and many dogs chew them beyond recognition. Also, if two dogs have the toy, they'll probably devise a more exciting game that doesn't include you.

Once a dog is hooked, introduce "game rules" one at a time, such as relinquishing the toy when requested, no bodily contact with you, and for peace of mind of neighbors, absolutely no barking. And don't forget: *Frequently integrate training into the game, so you may reinforce behaviors by resuming the game.* Ask your dog to Sit before you'll even open the toy box. Ask for multiple, short Heel-Sits and then a longer Sit-Stay before you throw. When playing fetch, ask for a Sit-Stay before you take the toy back, and then a few Come-Sits and a longer Sit-Stay before you throw again. Integrating frequent Sits into active games is the best way to teach impulse control.

When dogs become enamored with playing fetch or tug, balls and tug toys may be used as lures and rewards, especially for teaching rapid recalls and sparkly heeling with otherwise slow-moving dogs. And if your dog doesn't like balls, tugs, and Frisbees, teach them to looooove balls, tugs, and Frisbees by empowering the toys as mega-secondary reinforcers.

Empower Praise and Kibble as Mega-Secondary Reinforcers

Praise is intrinsically reinforcing; it is what's called a *primary* reinforcer. Additionally, dogs quickly learn that praise words are usually associated with, or immediately followed by, all sorts of cool things that they *really* enjoy, such as stroking, petting, dinner, couch time, a door being opened, and being let off-leash. As such, praise naturally becomes an extremely powerful *secondary* reinforcer, which predicts cool things are coming.

Let's say you get a formal letter in the mail. First you groan, but then, through the clear window, you see it looks like a check is inside, so you look for the return address and see it's from the lottery. Your heart races with excitement, and you rip open the envelope to find a check for $307 million. Wow! Now, the check is just a piece of paper with a little ink. Its actual value is less than a few cents. The check is a secondary reinforcer: It's not what it is (a piece of paper), but what it represents. What does it represent? Anything you've ever wanted!

Essentially, we can turn any stimulus or object into a secondary reinforcer that predicts better things, so let's start by empowering praise and kibble as *mega*-secondary reinforcers, so that each "thank you" and each piece of kibble becomes the doggy equivalent of a multi-million-dollar check.

Most people are familiar with clicker training, wherein a click — a neutral stimulus that has no prior relevance, meaning, or value to the dog — is repeatedly followed by a food treat — "click and treat." In time, the dog learns that the click predicts the treat, and so the click acquires the reinforcing power of the treat. We now call the click a secondary reinforcer because it predicts that food, a primary reinforcer, will follow.

Similarly, when desensitizing dogs to unfamiliar or scary stimuli, we repeatedly follow the scary stimulus with multiple food rewards. That helps make the stimulus less scary, and in a sense, the scary stimulus — such as an approaching person — now becomes a secondary reinforcer because they are the herald and harbinger of food rewards.

Praise and kibble are both primary reinforcers to varying degrees depending on the individual dog — some love both, some are indifferent to both, and others love one but not the other. However, few people think of turning these primary reinforcers into secondary reinforcers and so increase their reinforcing value. Regardless of their initial intrinsic value, give praise or kibble additional huge reinforcing powers as mega-secondary reinforcers.

When you integrate short Sits into massive primary reinforcers such as sniffs, walks, and dog-dog play, the act of sitting becomes a massive secondary reinforcer. So let's go one step further. Each time you ask your dog to Sit, take the collar, praise, offer kibble, and then say, "Go play":

Sit → Collar → Praise → Kibble → "Go play"

- "Go play" is a massive *primary reinforcer.*
- So, kibble becomes a *secondary reinforcer* predicting play.
- So, praise becomes a *tertiary reinforcer* predicting kibble and play.
- So, collar becomes a *quaternary reinforcer* predicting praise, kibble, and play.
- So, sitting becomes a *quinary reinforcer* predicting collar, praise, kibble, and play.

Both kibble and praise are still primary reinforcers in their own right, but as they become secondary and tertiary reinforcers, they acquire extra reinforcing power equivalent to play, which is one of the biggest primary reinforcers in a dog's world.

Use this sequence *every time you instruct your dog to Sit.* Integrate this into every aspect of your dog's life — on walks, at the dog park, playing games with you — and prior to *every* powerful primary reinforcer: before offering a stuffed chewtoy, opening a door, giving a chest scratch, belly rub, or piece of chicken, going on a car ride, and before you say the words *dinner*, *walkies*, *couch*, or *bed*.

Rewards are not what they seem; they are what dogs have been taught that they mean. Most observers just hear someone say, "Good dog," or see someone give a piece of kibble, and they think those things can't be very motivating. What they don't realize is that, through training, people have taught their dog to associate these two stimuli with everything in life that's good. They have amped up praise and kibble as *mega*-secondary reinforcers that predict everything a dog wants out of life.

Additionally, they have repetitively desensitized taking the dog's collar (the number-one subliminal bite trigger).

I also like to amp up tug toys as mega-secondary reinforcers when working with dog-dog reactivity. With reactive dogs, I don't like to use food or tennis balls to reward the dog in training if I don't know how other dogs might react toward the food or tennis ball. However, few unknown dogs react if you pull a tug out of your pocket and hold it up to your chest, so that your dog sits and focuses on the tug rather than eyeballing the other dog and escalating the encounter. To my dog, though, it's not just an indiscriminate tug toy; it's a mega-secondary reinforcer.

Teach Tricks

Teaching tricks is highly motivating and quickly transforms the entire atmosphere of training. It relaxes both people and dogs and changes their demeanor. Most people love teaching tricks, and most dogs love learning them because they make their people laugh and act happy.

Some people think tricks are frivolous or demeaning. I vividly remember an elderly gentleman in one of my puppy classes saying, "This is ridiculous. I didn't come to class to have fun!" Why shouldn't all training be fun? Whereas I don't think we should force people to squeak, giggle, and laugh, training can still be enjoyable for the dog, owner, and onlookers.

Teaching a trick is no different than teaching any other behavior. Just use the same four-step basic training sequence: (1) verbal request, (2) lure, (3) response, and (4) praise. However, there are considerable differences in people's approaches and attitude. When teaching obedience, people often give commands in a stern voice, the dog is expected to obey promptly with precision and attention, and praise is often missing or terse at best.

On the other hand, when teaching tricks, people give requests in an upbeat, happy voice; they often repeat requests without impatience or annoyance; and almost any response, successful or not, is met with giggles and laughter. Consequently, this inspires wiggles and wags in dogs, who learn to love tricks. For example, a person might instruct, "Rover... Bang! Bang! Bang!" They keep repeating the command until the dog eventually lies down in slow motion like an agonizing opera death scene, yet still apparently alive, tail wagging joyfully, and one ear cocked and one eye open to soak in audience appreciation.

Let's meld the best of both worlds. When teaching any cued behavior, give instructions in a clear, neutral tone, or use a handsignal, smile and praise when the dog responds, and reward handsomely when a dog responds promptly after a *single* instruction. All training should be this way. I get as much delight when watching precise, stylish heeling as when watching a dog beg or playbow on cue. But what I really love is seeing smiles and tail wags.

During a group first-dog-to-down competition in puppy class, a ten-year-old girl said, "Pooch, Bang!" (a single bullet), and her puppy dropped like a stone and remained motionless with their eyes closed. A man was still shouting, "Brooklyn, *Down*, *Down*, *Down*, *Down!*" while Brooklyn merely bounced to the beat. The man complained, "But her dog didn't lie down." The girl replied with the acerbic precision of Maggie Smith, "Bang!... is a *supine* down."

Aside from being enjoyable, tricks have many practical applications. For example, High five and Shake hands facilitate examining a dog's paws. Laugh or Smile helps to examine the teeth. Stand-Stay enables examination of the front, back, top, and sides of a dog's body, and Rollover or Bang! facilitates examining the dog's chest and abdomen. How else are you going to look at the dog's belly? With a mirror on a stick, like looking for car bombs? Or slide underneath on a skateboard? One enterprising veterinarian in Manhattan solved the problem of examining giant dogs (who hadn't been taught Rollover) by digging a hole in his examination room. Big dogs would stand on the floor next to the hole, while he went down five steps and looked up to examine the dog's undercarriage, like fixing a car in a garage.

Many tricks are a godsend during social encounters. So many tricks cue friendly and appeasement gestures, like Rollover, Beg, Shake hands, High five, Kiss the hand, Laugh, and Playbow. All of these present a much-less-threatening picture to unfamiliar dogs and people, which helps them feel more at ease with *your* dog, which in turn makes them less likely to present a threatening picture to your dog, which helps your dog feel more at ease.

Also, we know that if people act an emotion, they feel that emotion. If we smile or laugh, we feel happy inside. When dogs are instructed to display numerous friendly behaviors, maybe they, too, *feel* friendlier and more confident around people and other dogs.

As with teaching tricks, you'll soon learn that there's so much more than meets the eye when playing games with dogs; see "Playing Games," page 340.

The Jolly Routine

Way back in the seventies, industrial psychologist and dog trainer Bill Campbell championed a wacky procedure he called *the Jolly Routine*. Basically, when your dog encounters anything scary — another dog, a man in a hoodie, a skateboard or motorcycle, a flailing and screaming child, thunder, fireworks — you respond by laughing, joyfully clapping your hands in glee, singing, or dancing a jig. Becoming jolly immediately changes your affect and demeanor, which is infectious and changes your dog's attitude and feelings toward whatever triggered their fear.

The Jolly Routine was decades before its time and remains woefully underused today. Yet it offers a very effective and enjoyable means for preventing or resolving many doggy fears and phobias. The Jolly Routine fast-tracks classical conditioning and is also by far the best reward for a lightning-fast recall. I always used to dance a jig whenever little Oso pooped on cue, always backed uphill. So what if some passersby thought I was a little weird. My dog had just pooped on cue and was therefore joyfully empty. Shouldn't we both be jolly about that?

Be Silly: Sing and Dance

Never underestimate the power of silliness. No, this doesn't teach dogs anything in particular — except that their owner is a fun person to be with. Listen to music, sing, dance, talk, and laugh with your dog. Have chats on the couch, make jokes, goof around. Even be serious if you want: Discuss your work, your sports team, your family, or the state of humankind. But especially: Sing, dance, laugh, and waltz along the sidewalk. Now look at your dog. They will be looking at you, maybe with shock, maybe with shared enthusiasm.

I remember years ago, at a workshop, I was trying to motivate the men in class to lighten up and brighten up. They were grudgingly "praising" their dogs as if they were cowpats. So I got the six men in class to swap dogs with another person and form a line. I told them to follow me and copy everything that I did. I started to walk slowly around the room with my head bowed. We made for a sad picture — trudging along with bowed heads, embarrassed — and their borrowed dogs were equally unhappy and, I dare say, also embarrassed. I started to whistle a tune:

> Heigh-ho, heigh-ho,
> It's home from work we go.
> We work all day,
> And have no play,
> Heigh-ho, heigh-ho, heigh-ho, heigh-ho.

A couple of dogs looked up at the sound of the men whistling, and then, everything changed. It is simply impossible to walk and whistle that tune without sporting a jaunty step. Gradually, all seven of us dwarfs started to swing our heads and shoulders from side to side with each step, and one by one the dogs' heads came up, gazing quizzically, and then their tails started to wag, and they pranced proudly by the side of their jolly and joyful new male friends. Try it when you next walk your dog.

Internal Reinforcement and Self-Motivation

The ultimate goal of all this is for a dog's response to become its own reward. In other words, we want dogs to become internally reinforced and so self-motivated that "just doing it" is more than sufficient reinforcement. When that happens, the original four-step, basic training sequence becomes just two steps:

1. Verbal request → 2. Response

External rewards are no longer necessary. To use a human analogy: We don't *have to* pay children or adults to read, cook, walk, jog, ski, dance, or play golf. When taught successfully, all these activities become self-reinforcing, and so it is with the dog training game.

Initially, training requires effective reinforcement and ongoing guidance, but gradually, a dog's appropriate and desirable behaviors become habitual, and most praise and reward become unnecessary. Our dog develops *good* habits that no longer require reinforcing by us. Moreover, these good habits become set in stone because they are *continually self-reinforced from within.*

When powering up praise and kibble as mega-tertiary and mega-secondary reinforcers, we have already amped up the "act of sitting" as a quinary reinforcer. Now that our dog is becoming more motivated, we can accelerate the self-motivational process with much shorter sequences: Sit → "Let's go," Sit → "Go play," Sit → "Fetch," Sit → tug, Sit → "Couch," Sit → chest scratch, and so on. Because walks, play with dogs, games with us, and relaxing and luxuriating now consume so much of our lives together, eventually, the "act of sitting" becomes the mega-secondary reinforcer par excellence. As Phoenix used to say, *Done deal!*

Once dogs become self-motivated, you and your dog can get along with the important business of enjoying your lives together. You feel good and your dog yearns for your companionship. Closeness and playing together continually reaffirm the obvious, that your dog shares all your good feelings, too. This is the longed-for result of all successful training, and using life rewards is the only way to get there.

Nonetheless, I always continue to say *thank you* and chat with my dogs and offer many presents. I like to do it, and my dogs like me doing it. It's just that these rewards are no longer necessary.

The Secret of Reinforcement: Happy Hormones

Internal reinforcement is so incredibly powerful because it's fueled by a powerful pack of "happy hormones," including dopamine, serotonin, endorphins, and oxytocin. A healthy diet and regular exercise, especially enjoying habitual, repetitive, and focused activities, such as play/training with our dogs and walking outdoors in the sunshine, spur massive changes in endocrinology and neurotransmitters that exert wonderful effects on our brain, body, and behavior.

Technically, oxytocin is the only hormone of the four. It is released into the bloodstream from the pituitary and impacts many different organs and tissues in the body, including the brain. The other three chemicals are neurotransmitters that directly affect neurons in the brain. However, all four have wonderful effects on mood and feelings, and hence the way we act.

Dopamine is the queen bee in terms of internal reinforcement (within the brain). Dopamine is released during pleasurable activities, such as eating and play, and it produces feelings of well-being, satisfaction, and accomplishment. Most important, though, dopamine is released with the mere *anticipation* of upcoming rewards and pleasurable activities.

In terms of training, this means that the longer our dog waits for an anticipated reward, the more dopamine is released in the brain to reinforce stays, attention, and impulse control. This is another reason to keep food in your hand or pocket for as long as possible instead of quickly popping it in your dog's mouth after any remotely acceptable response. Plus, it's a good reason to prolong stays before releasing your dog to play or sniff. Let your dog *savor the thrill of anticipation.* Basically, dopamine reinforces a dog's desire to *want* to train and play with us.

Serotonin is a neurotransmitter that causes feelings of happiness, focus, and calm. It is well-known as a mood stabilizer that reduces anxiety and depression. Many antidepression drugs work by reducing the reuptake of serotonin in the brain so that the chemical stays around longer to work its mood-stabilizing magic. Serotonin is naturally boosted by any type of physical activity and exposure to sunlight.

Endorphins are released during stretching, exercise, and *laughter*, and they help to temporarily reduce pain and relieve stress. Endorphins interact with the brain's opiate receptors and cause a brief euphoria followed by a sense of relaxation. This is the source of the classic runner's high, or the sense of joy and accomplishment that follows a vigorous cardio workout, such as a power walk with your dog. Not my forte; I saunter. However, I would occasionally attack the infamous Tamalpais steps with a Malamute for draft.

Oxytocin is often called the "love hormone," since it plays a major role in parent-child bonding and most aspects of mating and childbirth. Oxytocin is also released by exercise, gazing, any kind of physical contact (touching, massage, hugging, cuddling), and conversation. Yes, praising and talking to our dog releases oxytocin, which promotes bonding, attachment, trust, empathy, and... looove.

Nearly all the activities and situations that release happy hormones in dogs do the same for humans. In other words, as training emphasizes life rewards and becomes internally reinforcing and self-motivating for dogs, it's doing the same with us. Doing our best to ensure that our dog is calm, content, and happy helps us become calm, content, and happy.

This all results in a most delightful interspecies positive feedback loop. When we see or sense that our dog is happy, when we see a tail wag, or dancing paws and rump wiggles, when our dog rolls over for a belly rub, or repetitively pushes their muzzle under our hand to bump-start stroking, we get a surge of oxytocin and dopamine. We smile and keep doing what makes our dog wag their tail. Deep relationships radically change hormones and neurotransmitters, which change mood and behavior in both parties.

In a sense, interacting with our dog is the safest means to naturally self-medicate. Training our dog, playing with our dog, walking our dog, and stroking and cuddling our dog are just so good for our heart, brain, and soul.

Chapter 10

Putting "Problem" Behaviors on Cue

Years ago in Scotland, during my usual postseminar question time, I was standing in the middle of a throng of people and heard a bit of a disturbance. I saw an elderly lady wearing a long flowery dress and her version of a Carmen Miranda hat. She was carrying a cane and batting people on the shins as she bustled her way toward me, repeating, "Excuse me, excuse me, I have a problem." (I should say so.) People let her through and then she exclaimed, "My dog barks!"

In a flash, I decided this was no time to be a smart aleck, and so I replied, "Ahh! Is that your dog in the crate at the back of the room?" She replied, "Yes. He's my show stud."

Her dog had been barking on and off throughout the seminar. Suddenly, she whirled around, strode over to the crate, kicked it, and screamed, "SHUT UP!" Her German shepherd backed up with flattened ears and went silent. People were shocked.

Now, there were three things wrong with what she did. Obviously, kicking the crate and screaming at her dog, but can you guess the third?

Gathering my thoughts and calming myself, I said, "You obviously have good control over your dog's behavior, but have you ever considered first shouting, 'Shut up!' and then you only need to go over and kick the crate if your dog continues to bark?"

The third wrong thing was that the lady didn't warn her dog prior to kicking the crate, so the dog was never going to learn to shush when asked, which meant the barking and crate-kicking would likely continue indefinitely. All she was teaching the poor dog was that he was living with an unpredictable crate-kicker.

The lady then completely shocked me by saying, "Oh good lord, yes. What a marvelous idea!" I could see the cogs in her brain turning. She actually got it!

At first, I thought she was a silly lady, but I was wrong. She was obviously very smart and interested, but she simply didn't know how to teach her dog to Shush on cue, or more likely, she didn't even know that was an option. So I said, "I have another suggestion. When you warn your dog, there's no need to raise your voice. Say, 'Shush,' very softly and hold a finger to your lips, and then kick the crate only if necessary."

This would resolve the shouting. She nodded her head furiously in agreement, so I decided to try to address the crate-kicking. "Also, there's no need to kick the crate. Your dog loves you but you're scaring him. If he barks in inappropriate settings, just make eye contact and gently admonish him, 'Rover... I asked *you* to *Shush*.' Then when he stops barking and lies down, praise him. Of course, all of this is much easier if your dog is on-leash right by your side."

She looked at me for a few seconds, eyes wide and starting to glisten. She thanked me politely and abruptly whirled round and walked away. She put her shepherd on-leash and was gone.

In the seminar the next morning, this lady sat with her dog by her side, petting him periodically. Her dog did woof a few times, but each time she would softly say, "Shush," put her finger to her lips, and then pet her dog the moment he stopped barking.

That evening, I held my usual Q&A session in the hotel bar, along with the other seminar attendees *and their dogs*. The lady had become a different person. Engaging. Smart. Scintillating. She had a flurry of questions for me, but before answering them, I said, "Your dog was spectacular today. You've done a great job, and I see that you've worked out the most important techniques on your own: to frequently praise your dog when not barking and to always praise when he stops barking. Well done!" She also got loads of praise from the other attendees, who only yesterday despised her. I explained that her next step was to lure-reward train her dog to Speak on cue, to facilitate proofing Shush. As I explained the process, everyone gathered around, including the bartender and other bar customers. In no time (social facilitation), we had a canine choir singing and shushing on cue.

Cuing Behavior "Problems"

Basically, there are three good reasons to teach dogs to "misbehave" on cue: (1) the dog's *peace of mind*, (2) our *convenience* for teaching them to cease and desist, and (3) to use the dog's favorite, albeit annoying-to-us, behaviors as the best rewards for teaching cease and desist.

When training, it is always smart (and kind) to consider your dog's point of view. Some people consider many dog behaviors — like barking, jumping up, playing keep away, and hyperactivity — to be annoying and problematic. But for dogs, these behaviors comprise all their favorite activities; most are normal, natural, and sometimes necessary behaviors.

Years ago, a "hydraulic model" was used to explain the expression of species-specific behaviors. Energy, visualized as fluid, progressively dripped into specific storage tanks to fuel each natural behavior. To prevent the tanks from filling up, the energy required periodic release; that is, the natural expression of species-specific behavior.

If release were prevented for one behavior, pressure would build in the system, and the energy would cause an explosion of that behavior, or it might be forced into other behavior tanks and so be expressed as other behaviors. For example, if a dog is denied exercise and play, or prevented from barking, the dog might find energy release elsewhere — by chewing their paws or chasing their tail instead. This is a very old behavior model, but I think it helps to visualize what's going on in a dog's brain, and often in our brains, for that matter.

So much of dog training has been devoted to trying to inhibit and eliminate natural behaviors. Suppressing behavior and trying to prevent a dog from acting like a dog just doesn't work too well, and it's not fair. All dogs love sniffing, playing, and wagging their tails, and they also enjoy barking, jumping, pulling, hyperactivity, chasing, and being chased.

Always, when we are addressing undesirable behavior, let's ask our perennial question: If this is "wrong," what is *right*? What would we like the dog to do instead? For almost every behavior that is expressed in a fashion that *we* consider to be undesirable or unacceptable, with just a little thought, we can come up with a few situations where the expression of the same behavior is both acceptable and desirable. For example, barking to let you know someone has just broken your car window in the driveway. Thank you, Dune. I only wish that I hadn't told you to Shush.

A mutually acceptable compromise between what *the dog* wants to do and what *we* would like the dog to do is obvious. If we establish verbal control over both the onset and offset of behaviors we don't normally appreciate, we can teach dogs the times, places, and situations for their appropriate expression.

How? First, by not waiting for the behavior to erupt. A dog barking like crazy as groups of costumed children knock on the front door during Halloween is not the best possible training scenario to teach a dog to Settle and Shush on cue. Instead, we choose a time of day and situation where we control all the distractions. After first relaxing with a nice cup of chamomile tea, we first teach the dog to Speak on cue, so that we may repetitively practice teaching and proofing Shush on cue *at our convenience.*

Here are a few of my favorite behavior "problems" to put on cue:

- **Elimination:** "Go pee" and "Go poop" in your toilet area
- **Chewing:** "Chewtoy," find and chew your chewtoy
- **Barking:** "Speak" versus "Shush"
- **Grabbing:** "Take it," versus "Off" or "Thank you"
- **Pulling on-leash:** "Hustle" versus "Steady," and "Pull" versus "Heel"

- **Running away:** "Tag" versus "Come"
- **Hyperactivity:** "Jazz up" versus "Settle down"
- **Jumping up:** "Hug" versus "Sit"

Eliminate on Cue

How do we inform our dog about their brand-new, fancy indoor or outdoor toilet area? Stake a sign that reads "Doggy Toilet"? Well, that might work if we wrote "toilet" in dog urine on the post, but there's a better way. We can teach our dog to eliminate on cue so that we may communicate *when* and *where* we would like them to pee and poop.

Housesoiling is a *temporal* problem as well as a *spatial* problem: Either a dog is in the wrong place at the right time (such as indoors when they need to go) or in the right place at the wrong time — such as outside with us in the rain at 11 p.m. because we think they need to go but our dog has already emptied their bladder beforehand on the bathroom carpet.

Successful housetraining depends on being able to accurately predict *when* a dog needs to go, so that we may *be there* to show *where*, instruct "Go pee" and "Go poop," handsomely *praise and reward*, and then *inspect and clean up*. A few repetitions of that sequence and a dog soon learns to eliminate on cue, which then makes life a whole lot easier.

Crate training with a potty break every hour is the easiest way to predict elimination and thus put the behavior on cue. Most dogs quickly fall asleep after a lengthy chewtoy chew. As they sleep, their bladder and bowels slowly fill, and so every time we wake them for a toilet break, they need to go. Another option is to keep your dog on-leash beside you with a food-stuffed chewtoy, and then you can wake your dog on the hour to eliminate.

Otherwise, we must be on vigilant watch for every time the dog stops playing, wanders off after a lengthy chew, or wakes from a nap, especially if the dog drank a lot of water after the last potty break. And we must be ready to spring into action.

Teaching a dog to eliminate on cue is so useful for when we're in a rush, when it's cold and rainy, before visiting friends' houses, during potty breaks when traveling, and when we need to potty our dog on public property.

Visiting Manhattan, I have always been fascinated by the local custom of dog owners transporting bags of doggy doo through the streets late in the evening. Why not instruct a dog to eliminate curbside right outside the apartment building? That way, people can clean up and deposit the bag in their apartment's trash can and then enjoy a poop-free stroll with their dog.

Last thing at night, we de-pee and de-poop five or six dogs in training, and once they are all done and back indoors, they all Sit-Stay in a semicircle for more praise, intelligent discussion, evaluations of their pee/poop performances... and kibble-time! My solo record is five dogs, in less than a minute, after midnight, in the rain. Beat that!

While in the doctoral program at UC Berkeley, I once researched sex differences in elimination postures of dogs (makes for spicy bar talk), and I have always enjoyed housetraining. I'm not joking. I remember coming home very late one night from the ballet (with Mimi) after Omaha's first evening on his own, and I must admit, I had fretted throughout the entire American Ballet Theatre performance. As soon as we got home, I rushed outside and instructed Omaha to urinate, "Omaha, *inua mguu*" (lift leg), and he immediately opened the floodgates. Then I instructed, "Omaha, *mbwa choo*!" and waited for him to poop. He sniffed and circled, and then sniffed and circled some more. And more. But he didn't poop. Mimi came out and asked why I was taking so long. I said, "I know he needs to poop but he won't." Mimi offered assistance. The two of us chanted the poop command over and over while circling the yard, "*Mbwa choo! Mbwa choo! Mbwa choo!*" Then... he pooped! We just cracked up. The thought of an Englishman and a Chinese woman trying to chant in Swahili to an Alaskan Malamute in Oakland after midnight, all to get him to defecate, was just too overwhelming. We collapsed in laughter, praised Omaha, and gave him half a dozen treats.

Chew and Settle on Cue

Teaching dogs to chew on cue and what to chew is usually insanely easy. Just give them a selection of chewtoys and say, "Here's a *chewtoy*!" Sometimes, though, providing chewtoys doesn't necessarily mean that the dog will chew them. I mean, consider what you would make of a rubber Kong if someone gave you one as a present. What on earth is this for? A plug for the bath? A sexual aid? A shock absorber from a motorcycle? Actually, shock absorbers were the epiphany for Kongs — the original dog chewtoys. Joe Markham (King Kong) was fixing his motorcycle when his dog grabbed a shock absorber and proceeded to bounce it all around the garden, pouncing on the shock and tossing and chewing. I think it was his mother who suggested that, with their unpredictable bounce and durability, rubber shocks would make ideal dog toys.

To spark your dog's interest in hollow chewtoys, fill them with kibble. If ever your dog is eyeing an inappropriate chewing target, or is already destroying one, simply instruct, "Chewtoy," and help your dog find one.

Stuffed chewtoys are long-term rewards because they lure dogs to settle down calmly and quietly with the toy, and then over the next thirty minutes or so, the dog is automatically and repetitively rewarded for lying down, not barking, and for chewing a chewtoy. Also, chewtoy-trained puppies rarely develop separation anxiety because they have become accustomed to numerous quiet moments on their own.

Teaching "Misbehavior": The Yo-Yo Technique

Elimination and chewing involve simply teaching the dog the appropriate expression of these behaviors. However, for other "problem" behaviors, like jumping up, we can troubleshoot them by teaching two cues: one to engage in the undesired behavior and one to cease and desist. We do this by yo-yoing between the two cues; for jumping up, that would be Hug and Sit. We toggle between the *onset* — what the dog would like to do, jump up and Hug — and the *offset* — or what we would like the dog to do, Sit.

Sit-Stay → Hug → Sit → Hug → Sit → Hug → Sit → Stay

Now, what we would like the dog to do (Sit) is repetitively reinforced by what the dog loves to do (Hug), and vice versa. Essentially, what we once perceived as a problem behavior becomes a very effective reward, one that can join the expanding list of life rewards that we use to reinforce training. For example, just as Hug reinforces Sit, a quick Hug can be used to reward a dog to reinforce a super-fast recall.

As another example, once we first teach our dog to Speak on cue, we may repeatedly practice teaching Shush at a time of *our* choosing and not when our dog is in the middle of a barking frenzy. Toggling back and forth between the two behaviors allows us to perfect reliable stimulus control of both Speak and Shush in a *single* troubleshooting session, so that in the future, we can ask our dog to Speak when it's OK or required and to Shush when it's not.

I consider "cuing behavior problems" to be a mutually beneficial, canine-human compromise that allows dogs to be dogs and provides them with acceptable opportunities to release the pressure buildup in their barking and hyperactivity tanks. We can cue occasional bark-athons or turbo-zoomy sessions, which we may end at any time, once we feel that our dog has released the pressure from their system. At that point, they are much more likely to be calmer and quieter throughout the rest of the day.

Barking on Cue

We cannot simply invite a dog into our home and insist that they *never bark in this town again*. That would be on par with punishing a canary for tweeting, a baby for babbling, or a spouse for "singing" in the shower. Instead, teaching our dog to Speak and Shush on cue allows us to indicate where, when, and for how long it's OK to bark.

For example, allow one yip to signal they would like to go outside (whenever *we* are not paying attention to their toilet needs), three or four barks to announce visitors at the front door, no barking in the yard even when people and dogs walk by, but an occasional one- or two-minute barkathon in the car to let off steam.

We trained Claude, our Big Red Dog, to only bark to signal FedEx, UPS, and USPS deliveries. Nothing used to irritate me more than finding a little sticker on the door letting me know that a delivery was attempted, but a signature was required, so the parcel was taken away to become irretrievably lost somewhere in the void between my house and their facility. With Claude's trained announcement, we never missed a delivery, bless his heart. When lumbering old Claude barked, Lois, my neighbor, would look out her window to check whether or not my car was there, and she'd pick up my delivery if it wasn't. Nowadays, of course, delivery personnel just chuck parcels over the gate, photograph the parcel, and drive off.

Dune was trained to bark only when nondelivery strangers were on the property and to not bark at anything else, including people and/or dogs walking by or squirrels trotting along the electrical wires.

The story of George in the Introduction is an example of teaching Speak and Shush on cue. Because barking is such a common problem, here's a summary of the procedure:

To teach Speak: (1) Say, "Rover, Speak!"; (2) have an accomplice knock on the door or ring the doorbell, or bark yourself; and when (3) your dog barks, (4) praise and maybe bark along.

To teach Shush: (1) Say, "Rover, Shush!"; (2) hold a piece of kibble for your dog to sniff; and when (3) your dog stops barking to sniff, (4) praise by counting the seconds — "Good shush one, good shush two, good shush three" — and offer a food reward. A reward only reinforces behaviors that occurred within the last three seconds, and so your initial praise might reinforce a couple of seconds of barking, but continued praise and the food reward reinforce *stopping barking* and *continued silence*.

With each Speak/Shush repetition, progressively increase the requisite silence required for a food reward. Once you get to ten seconds, you're practically there, since after ten seconds, most dogs have forgotten what set them off in the first place.

Video the training session. You want to capture your dog's facial expressions the first time you rejoice and join in with the barks: *They bark! They speak dog!*

Occasionally, when being lured to bark, especially when there's a camera nearby, many dogs will look quizzical and cutely cock their head to one side to listen to you *in silence*! This is Murphy's Law of Dog Training: "My dog barks at everything! My dog barks all the time!" Not today. Not when you're trying to teach Speak and Shush. I found on my TV show that the hardest thing to do was to get a behavior problem on camera. If you cannot show the problem, showing the "solution" is meaningless, and so the whole episode would be a bust.

Writing about teaching dogs to vocalize on cue brings back some lovely memories. Late one evening when driving home across the Bay Bridge from San Francisco, all lanes had come to a dead standstill. For three hours we were motionless. Our stomachs were growling and bladders screaming. I let Omaha pee against the rear wheel. I had to grin and bear it. Omaha was beginning to rock the car with his restlessness, and so I instructed him to howl. With his hind legs on the folded-down rear seats, he put his front paws on the center console, stuck his great dome through the sunroof, threw back his head, and let rip — a full-blasted, lingering plaintive howl. Attention and smiles were upon us, and then three cars over, a guy stood up through the sunroof of his BMW and howled in unison, beating his chest like a silverback. One by one, more people got out of their cars, or stuck their heads out of their car windows, and howled. A glorious howlathon, with a Malamute conductor, followed by giggles and calm. It was a truly wonderful and magical moment. Everybody venting, letting off steam, reducing the pressure buildup, and then collapsing in laughter. I laughed so much that I nearly wet my pants.

Years later, sitting on the balcony overlooking nothing but trees in Codornices Park in Berkeley, Big Red Claude let out a deep, soulful, hound howl, and after a couple of seconds, our dogs Ollie and Dune joined in, and then Kelly and I. In no time, our cheerful chorus set off other dogs in the hills and flatlands of Berkeley. After a few minutes, we instructed our dogs to Shush, and all was still once more. Thereafter, a communal howl with our dogs became a 6 p.m. wine-time routine. I have often wondered what our distant neighbors thought: *Who has those hounds in the hills? Why don't they shut them up? Can the dogs tell time?* After lovely little Ollie died, our Frenchie puppy, Hugo, *Le Petit Dauphin*, eagerly joined our pack. Not an auditory plus, though; he sounded like a wounded rabbit.

Grabbing on Cue

Some dogs grab everything: clothing, children's toys, cat toys, food on the counter, food in our hand, and even our hand, wrist, or arm. This is usually a sign dogs want to play, but it freaks out a lot of people, and so we must get it under control. Once we teach our dog Off / Take it / Thank you with nonsentient objects, we can use Off the next time our dog grabs us or a person.

Other dogs hold on and won't release objects they value. This can be a big problem. If this is an issue for you, consult with a dog trainer right away.

In chapter 6, I explain how to teach Off and Take it so that a dog doesn't bother the food lure (see pages 137–39). Teaching Take it (gently) with food requires no training. When practicing Off, simply say, "Take it," and open your hand, and the food reward is gone from your palm. The instruction Take it is also very useful for teaching dogs to reflexively take and hold other objects, such as tug toys, balls, Frisbees, and dumbbells.

Say, "Take it," and waggle the toy to make it come alive, repeatedly pull the toy back toward you to entice a dog to take the toy (without making skin contact). Once the dog takes the toy, praise and gently and rhythmically tug the toy to increase the dog's confidence to hold the toy.

Teaching a dog take and hold on cue makes it possible to repetitively practice teaching Thank you. This command means "Please give my toy back." Say, "Thank you," offer (but don't give) a piece of kibble with one hand, hold the other hand under your dog's muzzle to catch the toy, and then when they release the toy, give the kibble as a reward. Repeat the Off / Take it / Thank you sequence over and over.

When playing fetch, you want your dog to sit and deliver the object into your hand. With the game tug, you simply want your dog to release the toy and sit while you keep hold of the toy. When playing tug, I hold the tug at all times; say, "Thank you"; then I go perfectly still, apart from offering food with the other hand. As soon as the dog opens their mouth to take the food, I pull back the toy and hold it behind my back, and then I offer additional treats.

Eventually, a dog learns that relinquishing the toy doesn't mean forever, and after eating the kibble, they will get the toy back. This *token system* gives a dog confidence to let go when requested. The Off / Take it / Thank you sequence teaches three words that have so many uses.

With dogs that I don't know, I like to start training Thank you with a tug toy because the dog quickly becomes comfortable with me holding an object that they are holding in their jaws. I repeat Sit / Off / Take it / Hold / Tug / Thank you over and over. Then I repeatedly practice Off / Take it / Thank you with many other low-value objects, such as a rolled newspaper, dumbbell, book, stuffed animal, and chewtoys. Then I move on to toilet paper rolls and sterilized bones, and I do all this before I even think of working with a dog that guards meaty bones or toilet paper.

Dogs perceive the same object — such as a bone or a piece of toilet paper — very differently depending on whether we are already holding it or reaching for it. In our dog hierarchy studies, we found that a bone on the ground can often be up for grabs, but if another dog has a single paw on the bone, it's not. And it's especially not if the other dog is female and grasping the bone with two paws while dismantling it in their jaws. Females are much more possessive of objects than males. In fact, even a very low-ranking female can keep a high-ranking male from taking her bone, simply by hunkering down and holding on to it. A female's dogged tenacity can crack a male's more fragile temperament. For female dogs, their First Amendment to Male Hierarchical

Law is: *I have it, and you don't!* Their Second Amendment: *You have it, and I want it.*

As I write, I often overhear imaginary dialogues of dogs discussing my training techniques. Here, I thought of a young female puppy conversing with a wise old Akita. The puppy says:

I'm confused and really worried about my person. Just this morning, I was chomping on this dry old bone and my person comes up and politely says, "Thank you," while offering me a little piece of steak, cooked but rare, plus a piece of freeze-dried liver. So I took the food and let him have the dirty old bone. A no-brainer, right? But here's what worries me. He gave me back the bone! I mean, he would never survive in a dog pack. Surrendering a bone!? The old geezer is losing his mind, and it gets weirder. The codger had squeezed a glob of peanut butter inside the bone! Beyond yummy, by the way. And then, minutes later, he did the whole thing again: said, "Thank you," and exchanged the bone for steak and liver. Then fifteen minutes later, he did it again! I think he's going off his rocker. He cracks me up, but I still love him.

The Akita replies: *You know, humans are quite inscrutable. I'm sure your person means well, but people do so remind me of cats.*

Pulling on Cue

You may not want to go the whole hog and teach your dog to pull on-leash on cue (although I do), but I think everyone needs to teach their dog Hustle (speed up) and Steady (slow down) to facilitate leash walking. By now, the formula should be familiar:

First, teach your dog to accelerate on cue: (1) Say, "Rover, Hustle"; (2) immediately walk faster; and when (3) your dog speeds up to catch up, (4) praise and maybe reward.

Then, teach your dog to slow down on cue: (1) Say, "Rover, Steady"; (2) immediately walk slower; and when (3) your dog slows down to walk by your side, (4) praise and maybe reward.

Yo-yo Hustle and Steady and practice on every walk, on-leash and off-leash. Walk using three gears — slow, normal, and fast — and frequently change up or down through the gears by instructing Steady or Hustle. If your dog doesn't slow down sufficiently, repeat, "Steady," and slow down even more. Or stop and instruct, "Heel," so that your dog comes and sits by your side. Similarly, if your dog continues to lag, repeat, "Hustle," and abruptly speed up again. If necessary, repeat, "Hustle," once more and change to warp drive.

Using pulling on-leash as a reward is a big one. To do so, first teach your dog to Hustle and Steady on cue, so you can yo-yo leash pulling with walking by your side. However, a surprising number of dogs just watch in disbelief when you give them the opportunity to pull on cue. Murphy's Law again.

So teach your dog the same way as I taught Blackie the heifer with the help of my cousin Paul and a cabbage. Enlist an assistant to walk in front of you with a lure to waggle, such as food, a tennis ball, or a tug toy. Tell your assistant to follow the directions you give to your dog. When you say, "Rover, Pull," the assistant should move away faster than you're walking, and your dog will pull to follow the pace of the lure.

The very first session that I taught Omaha to Pull on cue to cure his minor but persistent heeling habit of "showing me the way," he got it. His heeling improved that night as we merrily alternated pulling and heeling all the way to Bushrod Park. Pulling on cue has a number of uses.

Omaha and I once won a Northern California Malamute carting race together. About fifteen contestants were in a long line at the start, all of us in sleeping bags. The challenge was to get out of our sleeping bag in our long johns, roll the bag and put it on the cart, dress, and then mush forty yards to the finish. Omaha won, not just because he mushed when I said, "Pull," but mainly because he, unlike most of the other competitors, went in a straight line. He knew directions well because we had been perfecting a new sport, Malamute-propelled sled tree-slalom. Actually, truth be told, we never perfected the sport. But at least, with Omaha showing me the way, cross-country skiing became a noncardio event.

Every Sunday, during our two-family, Dunbars-and-Rowells hike up the Big Springs Trail in Tilden, Omaha came along already kitted out in his sledding harness. He would tow Jamie's pram up the rocky slope. Once Jamie was older and insisted on walking, Omaha pulled me up the slope. Like Jamie, I too had gotten older.

Running Away on Cue

Dogs thoroughly enjoy chasing and being chased. If you can teach your dog to run away on cue, it becomes easy to yo-yo running away and recalls, and unreliable recalls are no longer an issue. I call this game Tag. A lot of dogs get it right away and zoom off the moment you say, "Tag." Give chase for a while and then say, "Come," and run away from your dog. This is a game that dogs frequently play with other dogs.

However, other dogs are reticent to start the game. Most often because they are uncertain. Teaching a dog to run away on cue requires a lot of dog savvy. You must read your dog carefully to make sure they are not becoming stressed.

I usually train in slow motion until the dog gains confidence. I start indoors by seeing if I can get the dog to walk in front of me around the house, from room to room and around the furniture. Whenever we come to a door or a tight gap between furniture, I say, "Go on, Hustle," and gently shoo the dog in front of me while praising lots — especially if they speed up to shoot through a narrow gap.

Then we move to the garden or a fenced park, where I pretend I'm a sheepdog driving sheep. I do it all in slow motion to keep the dog calm, but also because I am *not* a sheepdog. Once the dog is ahead of me, I try to drive the dog, and I especially try to get the dog to change directions by walking forward a little to the right or the left of the dog. Then I turn around and move away from the dog and often hide behind trees, the same way a duck-tolling retriever entices ducks to swim within gun range. I yo-yo herding and tolling, and once the dog gets the idea, I instruct, "Hustle," to get the dog to move quicker when walking away. At this point, I repeat the Hustle command with a little more urging, so that I can alternate running away with speedy recalls. Once dogs have learned to run away on cue, I give careful chase.

Because of the more dignified pace, I've come to prefer herding/ tolling to Tag/Come. I always hurt the day following a good game of Tag, or Tug, or even after throwing a ball for ninety minutes. Herding really improves your dog savvy, and tolling is just pure fascination, especially when you start hiding behind bushes.

Tag was Omaha's favorite game. Following a Sit-Stay, I would say, "Tag," he would run away, and I would give chase. If I could grab his tail, game over. Every minute or so, I asked for a quick Come-Sit, and then we'd play Tag again. Physically, Omaha was uncatchable. I never managed to grab his tail even though our lawn was tiny. He would run and whirl and twirl and wrap his tail between his legs whenever I got close. But his normally leisurely recalls became quicker and quicker.

With Dune, whenever I said, "Tag," off he would go like a rocket. We played indoors, and he would run all over the house, upstairs, downstairs, and around the couches in the drawing room. But we would inevitably end up running around and around the dining room table, which he would circle so fast that he would catch up and chase me. Then I'd abruptly change directions and off we went again. Great rainy-day, indoor activity.

Hyperactivity on Cue

Obviously, just as dogs are dogs, children are children, and both have a limitless capacity for creating noise and activity, which can sometimes jangle the nerves. For myself, it's not so much the noise and activity but rather the *unpredictability* of the outbursts. I first played this game with Jamie. We put noise and activity on cue — *silly time*. And what a marvelously fun and effective game it turned out to be. Before visiting relatives, going into restaurants, movies, the doctor's office, or boarding a plane, I would explain to Jamie what was going to happen and how long it would take, and I would ask *how much* silly time he needed. Then we found a secluded setting so other people wouldn't freak out, and I'd say, "Jamie, be silly!" And I would join in. Once depressurized, either I or Jamie would signal, "Normal Jamie."

When putting "problem" behaviors on cue, we are cuing both the onset *and the offset.* This allows us to practice the cease-and-desist cue many times over. These cues are a boon for real-life scenarios. When Jamie and I practiced at home, Omaha was utterly fascinated and obviously wanted to join in. So the Jazz up / Settle down game for dogs was born.

The game came to fame after I gave a training workshop at an enormous pet dog event near Windsor, England, organized by Beverley Cuddy of *Dogs Today*. When I arrived at the arena, a good eighty dogs were already there, all of them joyfully barking and jumping as their owners desperately held on to their leashes. The noise was deafening. So many of the dogs' owners were fans of my television show, and I think they were expecting some sort of blessing or laying on of hands to make their dogs whole. The dogs were so out of control, and I had absolutely no idea what to do ... until Omaha spoke to me from the heavens, *Dad, tell them to Jazz up!* I plunged into the deep end.

I told everyone, "When I say, 'Jazz up your dogs,' I want everyone to get their dogs to leave the ground and vocalize, until I raise both my arms and say, 'Settle your dogs,' whereupon *everyone* must get their dogs to *settle and shush*. We'll repeat this over and over until *everybody* can settle their dogs from Jazz-up mode in *less than three seconds*." A tall task indeed. I had never played this game with as many as eighty dogs, but I've never lacked for optimism and confidence. I pulled out my stopwatch and said, "Jazz up your dogs!"

Everybody started making weird noises, waving their arms, hopping about like coked-up frogs doing John Cleese funny-walk impressions. As for the dogs ... Murphy's Law of Dog Training kicked in, and few responded. Many were frozen in place. Some looked dumbfounded, others looked positively shocked. The only appropriate word for the expression of one German shepherd was disdain: *Was ist los? Du bringst mich in Verlegenheit!* So I taught everyone how to get their dogs amped up. It has always amused me that the quickest way to stop a dog or child from being silly is to be sillier yourself.

The first few Settle downs took a good ten minutes before seventy-seven dogs were lying down calmly and quietly, but after a dozen Jazz up / Settle down repetitions, seventy-seven dogs met the criterion.

With the exception of very short breaks to catch their breath, the three other dogs barked and bounced for the duration.

What was charming, though, was that the other dogs' owners and members of the audience were all shouting encouragement to the owners of the three barky-bouncy dogs.

So off you go! Try this yourself. Practice yo-yoing Jazz up and Settle down until you can get your dog to settle and shush within three seconds. I call troubleshooting sessions like this *meet the beast*. Yes, it is going to be difficult, but because you are now focused on solving the problem, you *are* going to conquer the beast.

Jumping Up on Cue

Of all the "misbehaviors" to put on cue, jumping up has to be the simplest to solve. I found it so hard to stomach most of the recommendations in older dog training books about how to stop dogs from jumping up: to shout; to squirt dogs in the face with water, lemon juice, or vinegar; to squeeze their front paws, stamp on their hind paws, or hit them on the head with a rolled-up newspaper (UK version); to hang them with the leash; and to grab them by the scruff and flip them on their backs and hold them down in an alpha rollover! All of this just to stop a dog from enthusiastically jumping up to say hello.

Jumping up to greet people is pretty much the default and accepted greeting for most puppies and is often met with laughter and vigorous petting, which of course unintentionally reinforces the behavior. The puppies' only crime? *They grew!*

All of this when a simple Sit is the obvious solution. To increase response reliability, invite a bunch of people over for a hugathon: Have every visitor yo-yo Hug and Sit.

Several times in my life, I have been on some exhaustive, seminar trails. For most of the nineties, from February to November, every week I used to fly to give three- and four-day seminars. I would get back Monday mornings at 2 a.m. It was exhausting.

But every time I got back home and unlocked the front door, Phoenix was sitting in the hallway, waiting to greet me. She knew the routine, and it was quite comical. I would take off my lecture suit jacket and then my trousers and hang them on the back of the front door. (Greet a Malamute and you look like a Yeti.) Phoenix waited patiently until I said, "Give us a hug, Phoenie." Then she would waddle-wag toward me, raise herself on her hind legs, gently plant her front forelegs on my shoulders, and nuzzle my ear as I hugged her huge hairy body. Home! At last.

I'm not daft. I was savvy to her real motives. After the hug, we'd go to the kitchen. I'd pour a glass of red wine for myself and ... *grab a few biscuits for Phoenix*. And why not? *Phoenix was there* to give me a welcome home hug when everyone else in the house was sound asleep.

Sometimes a doggy hug is exactly what we need.

Part 4

LURE-REWARD TRAINING STAGE THREE: COMPLIANCE

I think we've been making a mountain out of a molehill regarding misbehavior and noncompliance. Just like people, no dog is perfect and they occasionally make mistakes, which is hardly surprising while they're still learning. Offering additional verbal guidance is a game-changer for resolving misbehavior and noncompliance.

The dog's ever-expanding vocabulary facilitates resolving misbehavior, and in most instances, just a single-word instruction quickly gets dogs back on track, especially when we troubleshoot the most common scenarios that trigger difficult-to-control hyperactivity and noise, such as visitors, encountering strangers and other dogs, car trips, and so on.

Similarly, verbal feedback and guidance is a game changer for dealing with noncompliance. With calm and persistent insistence, dogs progressively *learn the relevance* of our instructions and response reliability further improves dramatically when we use two names for our dog: a *formal name* to clearly signal formal instructions and to let dogs know when they are "on duty" and compliance is necessary, and a *nickname* to let dogs know that they are "off-duty," and so for the most part they can chill and relax.

As training progresses, learning distance and emergency commands vastly improves a dog's quality of life by allowing dogs considerably more supervised freedom to explore, run, and play off-leash.

Chapter 11

Single-Word Instructions

I teach dogs ESL so I can give clear *verbal* instructions and ongoing *verbal* guidance as they progressively learn. However, the old *binary feedback* of reward versus punishment remains chipped in the stone of our brains and many people still feel: If we reward a dog for desirable behavior, we must punish a dog for undesirable behavior. Let's examine how well that works.

Do We Reward, Punish, Ignore, or Inform?

In seminars, in order to illustrate the relative effectiveness of the most common combinations of feedback people offer to dogs, I would play my version of Karen Pryor's "training game." I would ask for six volunteers to be human-dogs, and I'd send those people out of the room, while I would write the task to be taught on the whiteboard. My standard task was: "Fetch Joel's beer from the back of the room and bring it to Ian." Then I would ask for volunteer trainers to administer the various permutations of feedback for desirable and undesirable behaviors.

I have always thought that *how a dog perceives training* is a massive consideration when training, and this game gives everyone a pretty good idea of how dogs probably experience praise and reprimands. Instead of using kibble and physical correction, I suggested the volunteer trainers use verbal feedback, for example, "Goooood human" for praise and "Baaaad human" for reprimands, so that the game resembles playing physical charades with "hot-and-cold" feedback.

1. Praise *for Correct Responses and* Reprimand *for Incorrect Responses*

The speed of learning usually depended upon the ratio of praise to reprimands, plus the nature of initial feedback. With some initial praise, the human-dogs would try more options, seeking to figure out what was wanted, and so long as praise continued, they learned quickly. But after just a couple of reprimands, especially when delivered in rapid succession, many volunteers would become inhibited and move cautiously, until learning came to a standstill.

2. Ignore *Correct Responses and* Reprimand *for Incorrect Responses*

From the early 1900s until the 1980s, this was the predominantly accepted style of feedback in dog training, with the exception that dogs were *physically* corrected. The formalized on-leash training of military dogs was later adopted and adapted for on-leash competition and working dogs, and eventually for companion dog classes. Next to no (comprehensible) instructions were given to dogs prior to the task, and feedback focused on physically "correcting" *all* misbehavior and noncompliance, while ignoring most desirable behavior. In the training game, after just one or two reprimands — "BAAAD human!" — every human-dog froze and gave up.

3. Ignore *Correct Responses and* Ignore *Incorrect Responses*

This option may sound daft because, at first glance, it would seem to communicate nothing. But lack of feedback communicates a lot: *Do what you like.* Lack of feedback epitomizes what so many people offer their dogs: They fail to praise or reward much, and they fail to offer much guidance when their dog goes off track. Also, this is the only type of feedback trainers are allowed to give dogs in competitive obedience — no feedback. Dogs are required to respond to a single verbal cue or handsignal with no feedback whatsoever.

In the training game, most human-dogs would become embarrassed, impatient, and exasperated, and eventually they would ask, "What am I meant to do?" Precisely.

4. Praise *for Correct Responses and* Ignore *Incorrect Responses*

As in wait-and-reward training and shaping, the dog is rewarded for desirable behavior, and all unwanted behavior is simply ignored. The dog is given no instructions prior to the task, and so noncompliance is not an issue. If the behavior suddenly appears in its entirety, the dog is rewarded. With shaping, the dog is rewarded for progressive approximations to the final performance, with training proceeding in a stepwise fashion.

In the training game, I asked for an experienced clicker trainer from the audience to use their clicker and pretend treats to train the human-dog. However, shaping an adult human is much more difficult than shaping animals and children because people often overthink and ponder *why* after every click. Like dogs, our human-dogs would often get stuck. For example, when clicked for taking a step in the right direction at the same time as passing a person, the human-dog would characteristically attend to the person and try all sorts of irrelevant and incorrect responses. They might touch or prod the person, or pick up every item on the table, and so the human-dog might go a long time without a click.

5. Praise *for Correct Responses and Give a* Nonreward Mark *after Incorrect Responses*

A nonreward mark is any neutral signal that conveys: Your last response is not rewarded. Trainers were allowed to choose between "nope," "try again," or "incorrect." A nonreward mark is not direct instruction, but it provides very valuable information: "You haven't made a mistake, but you made an incorrect choice, so keep trying and have another go." When teaching dogs, there is no reason to use a loud or unfriendly tone of voice to inform dogs that they have made an incorrect choice; the information alone is valuable.

In the training game, nonreward marks were highly effectively at getting a human-dog out of a rut. They helped training proceed much faster than when incorrect responses were ignored by the trainer.

6. Praise *for Correct Responses and* Instruct *after Incorrect Responses*

As fun as the training game always was, our human-dogs grew impatient for the "correct answer," and so to speed up the process, I took over to train the last human-dog. I'd say, "Rover, Go to . . . Joel. Good human. Gooood human." Then, once they reached Joel, I'd say, "Rover, Fetch beer. Come here. Thank you. Goooooood human." Then I would reach under the podium to get a beer for them, open both beers with my lighter, and we would toast, "To verbal instructions!"

Lots of people in the audience would cry foul: "You're cheating!" I would reply, "No, I'm not. This is exactly how I would instruct a puppy to perform a new task. In this game, I only used two words that a puppy wouldn't understand after Puppy 1: *human* and *beer*, and in puppy class, I would say 'good dog' instead of 'good human.' We don't teach *beer* until Puppy 2." As it happens, a lot of men in puppy classes love to teach their dogs, "Rover, go to the fridge and fetch a beer." This appears to be a strongly gender-specific desire.

By playing the training game, people got to experience (some directly) what it's like to walk in a dog's paws. The game quickly drove home the limitations of noninstructive reprimands and the power of praise and verbal instruction. The takeaway:

- Not giving clear, comprehensible instructions beforehand seriously handicaps learning. In fact, in the sixth scenario, I barely gave any feedback. My initial instruction was sufficient, as it often is with dogs.
- A high praise-to-reprimand ratio is critical for initiating and maintaining engagement.
- Praise works exceptionally well by itself.
- Training slows down without verbal guidance when dogs make wrong choices. Nonreward marks speed things up, but verbal

guidance provides the solution. As we played the game, I had to frequently shush members of the audience from whispering suggestions and clues.

Single-Word Instructions for Misbehavior

During the last five years of my lecturing career, I always liked to get to the nitty-gritty and ask seminar attendees, "Can you suggest non-aversive solutions for misbehavior and noncompliance?"

For misbehavior, many suggested turning your back on the dog and waiting for the behavior to extinguish. Others suggested distracting the dog, interrupting the behavior, teaching another behavior, and especially, teaching an incompatible behavior.

I would be much more specific. Let's go back to our basic question: If this is "wrong," what is "right"? How would you like the dog to act *right now*? Obvious, right? So let's just inform the dog. If a dog is barking, instruct, "Shush." If a dog is chewing your shoes, instruct, "Chewtoy." If a dog is careening around the kitchen, instruct, "Bed." If a muddy-pawed dog is running upstairs, instruct, "Downstairs" or "Outside."

For training to be quick and effective, we need to use words. We need to use clear instructions prior to the task, offer words of praise when dogs make correct choices, and when dogs make incorrect choices or "misbehave," we need to concisely and precisely, often with a single word, instruct dogs what to do. Single-word instructions convey two essential pieces of information:

1. Stop what you're doing.
2. *Do this instead.*

For everything that dogs do that remotely gets our goat, we need to ask ourselves: If this is not what we want, what is? Then we simply tell our dog *how* to act. Why a single word? To keep it as simple as possible for the dog.

As you may have noticed, I have been doing this throughout the book. Sit is the solution for jumping up and so many other misbehaviors; Outside or Toilet for when a dog is about to eliminate indoors; Thank you, Off, and Chewtoy for when a dog is chewing or eyeing an inappropriate object; Steady when dogs are pulling on-leash, and so on.

Some words have multiple applications. Off can mean don't touch my computer, eyeglasses, cup of coffee, piles of tax return receipts, or the food in my hand or on the counter.

The commands Sit, Down, and Shush all help dogs to calm down when they are stressed and anxious and about to bark themselves into a frenzy. Depending on the situation, other useful single-word commands for overamped dogs are Settle, Steady, Chewtoy, and Bed.

Further, it can work well to give a string of single-word commands, so that they almost resemble a sentence. For example, if a dog already has an inappropriate object in their mouth, you might say, "Thank you, Chewtoy, Bed, Settle." In fact, I love giving instructions in a sentence, emphasizing the words that the dog understands. Once we've taught our dog the meaning of all these words, let's use them to help our dog navigate life.

Sit is the King of Commands because it has so many uses. It immediately arrests so many frustrating, annoying, and sometimes dangerous behaviors: dashing through doors and gates, running away, galloping around the house, chasing the cat, jumping on visitors, over-the-top play with people and dogs, and more. If you are ever in doubt, just say, "Sit."

Using single-word instructions to eliminate misbehavior is pure simplicity *in theory*, and when dogs are calm, *in practice*, too. However, there are occasions when a dog is so overwhelmed with excitement, or so focused on a triggering event, that they barely acknowledge that their owner is on the same planet, let alone trying to offer useful guidance. At times such as these, the dog's brain simply cannot receive instructions.

You probably already know most of the scenarios that set your dog off. Don't fret or fear, but the time is nigh for you to meet the beast and *troubleshoot crazy*.

Troubleshooting Crazy

It's time to focus on those specific situations where a dog appears to lose their mind and their hyperactive rambunctiousness and rumbustiousness go way over the top and become uncontrollable.

Most owners try their best to control their dog. They say, "Sit! Quiet! Settle down!" over and over, but their dog is completely unresponsive. At this point, many people give up on training and try their best to hold on to their dog's collar or leash or resort to other measures. They might try to avoid triggering events by confining their dog in another room when visitors arrive, no longer inviting visitors, no longer going on car rides or walking the dog, or only walking the dog after 2 a.m. Other owners just let crazy happen and grin and bear it.

The real problem is that people often choose the wrong time to train. Trying to teach a dog when they have already lost their mind is a next-to-impossible training scenario. Instead of trying to deal with the problem as it is happening during the course of everyday life, and while we are swamped in mayhem, set up a troubleshooting training session specifically to resolve the problem, so we may devote our full attention to training.

First, conduct a little prior preparation: Test your dog's response-reliability percentages for Sit, Down, and Shush in the most common places where your dog often "loses it" — for example, the front door hallway, the car, and on the sidewalk — but with as few distractions as possible. Make sure that your dog has a good understanding of how you would like them to act in these settings. Also, when you use human assistants with training (see below), make sure they have been instructed how to act around your dog.

Then you're ready to troubleshoot crazy, not by *avoiding* the triggering stimuli, but by *desensitizing your dog to the triggering stimuli with repeated exposure.*

Basically, when dogs are exposed to an exciting stimulus, for example, visitors, the first exposure will incite the biggest reaction — often an exaggerated and overblown reaction. And this is how people live with their dogs. Every visitor triggers an overblown reaction and after a prolonged wait for the dog to eventually calm down, life continues. Until the next visitor and the next overblown reaction.

The most effective time to train your dog is right after they have calmed down. The initial exposure to any triggering stimulus is always the worst, but thereafter, subsequent exposures cause progressively lesser reactions. Consequently, we can troubleshoot problems in a single session by repetitively exposing the dog to the triggering stimulus or situation over and over, until the dog barely reacts at all. In this way, the dog regains focus and impulse control, and so becomes more likely to attend and respond to our instructions.

As a human analogy, think of the opening sequence of the movie *Love Actually*, in which multiple people greet in the arrivals lounge at Heathrow. After not seeing each other for weeks, months, or years, when greeting their loved ones, some of the people are beside themselves with excitement and hug, kiss, and in some cases, climb all over each other. The excitement continues as they travel home together and catch up over dinner. But fast-forward to the next morning.... Does the new arrival get the same emotional greeting after coming out of the bathroom?

There are plenty of good candidates for *troubleshooting*: the two happiest times in a dog's day, dinner and walkies; going through doors and gates; car rides; greeting visitors; encountering people and dogs on walks; and going to dog parks. Even something as simple as putting on a dog's leash can become an extremely time-consuming rigmarole if a dog can't stop bouncing with excitement.

Since the troubleshooting technique is similar for all scenarios, I'm going to focus on four of the most common and challenging scenarios: car rides, greeting visitors at home, on-leash walking, and greeting people and dogs on walks.

Car Rides

Here's the deal: Just because your dog responds reliably in the house, and even on walks, doesn't necessarily mean that your dog will be well-behaved in the car. To get your dog to respond reliably in the car, you need to train your dog in the car. Moreover, one cannot drive a car and train a dog at the same time. First train your dog in the car when the car is stationary, and then ask a family member or friend to drive while you ride shotgun and continue to train your dog.

Since this is a new exercise that teaches some new commands, use lots of food rewards during the first session, so your dog has a good first impression of car rides.

To start, while outside the car with the door open, yo-yo between the instructions "Jump in / Lie down" with "Sit / Jump out." Repeat a few times. Every time your dog jumps out of the car, quickly ask for a Sit-Stay.

If your dog is reticent about getting in the car, it helps to have both back doors open and a person at each opening to call through the back of the car — Jump in, walk across the back seat, Jump out, and Sit. And then have the other person call the dog back in the other direction. Yo-yo the direction numerous times. No animal likes to enter a darkened space.

Then, while your dog is in the car, practice basic position changes and stays, and preferentially praise and reward your dog when lying down. Teach your dog specifically *where* to settle, for example, "back seat" or "crate." Troubleshoot putting on and taking off your dog's seat belt. Also, yo-yo Speak and Shush.

Sit in the car, turn on the radio, start the engine, maybe run the wipers and honk the horn, and then sit a while and check your texts, but all the time keeping an eye on your dog in the rear-view mirror, praising lots, periodically handfeeding kibble, and reinstructing when necessary. The deal is, if your dog doesn't lie down calmly when the car is stationary and you're paying full attention, your dog certainly won't lie down calmly when you're driving and not paying attention. Prior proper preparation!

If you're on private property, you could back up the car a few yards, then stop, turn off the car, and practice unloading. Then immediately ask your dog to jump back in and settle, turn on the engine, drive forward a few yards, and repark. Praise your dog lots as they remain settled and quiet. Keep your eyes and ears on your dog all the time in the rearview mirror. If your dog barks or moves location, respond with an immediate Shush or Down.

Now you're ready for a chauffeur. Have your driver drive around the block and park outside your house. Then you and the dog should get out, go indoors, and immediately come back out for another lap. Then have your driver take you both on a longer ride while you watch and praise your dog.

Now it's time to switch roles; you drive and your assistant can watch the dog. Periodically check your dog in the rearview mirror, praise lots, and offer other verbal guidance if necessary. If your dog has difficulty with any aspect of the car trip, for example, barking or whining, troubleshoot that separately.

Greeting Visitors at Home

The "nice-in-theory-but-doesn't-work-in-practice" dilemma often raises its ugly head when it comes to visitors. But by now you might guess the plan — throw a greeting party! The way to troubleshoot greeting visitors is to practice multiple repeated entries with the same person or people in the same session.

First, check response-reliability percentages when randomizing Sit, Down, and Stand with variable length stays. Then do the same with Speak and Shush in several rooms of the house, but especially in the front hallway on a mat or bed six feet from the front door. For front-door greetings, it is better to have your dog lying down behind you to prevent your dog from rushing the door when it's opened to let visitors inside. You don't want your dog going outside while an unknown visitor comes inside.

Before throwing a troubleshooting party, practice with a family member or friend — who has already been informed of the routine. Once the "visitor" knocks on the door or rings the bell, your dog will likely go ballistic. However, this time is different; this is a setup, and so there's no hurry for you to open the door. Consequently, devote your full attention to calmly and persistently instructing your dog, "Rover, Bed. Down. Shush. Good dog." It doesn't matter how long it takes or how many times you need to reinstruct. This is just your first attempt, and your dog is very excited. Things will only improve from here on because today you're facing the problem and going to resolve it.

Only use verbal instructions. Do not put your hands on your dog or their collar. No need to raise your voice, either; "keep calm and carry on." Remember, your dog will magnify and reflect your demeanor. Keep insisting and praise your dog lots: "Good Down. Oh goood boy. Shush." If your dog barks, say, "Rover, Shush." If your dog breaks the Down, immediately reinstruct, "Rover, Down." If your dog moves from their bed, immediately instruct, "Rover, Down." Praise until your dog is solid once more, and then reinstruct, "Rover, Bed. Down," and begin praising again. Again, there's no hurry to open the front door; the "visitor" has their instructions.

Once your dog is settled and quiet, say, "Enter." This cues the visitor to open the door a crack to peek inside and then close the door. (Don't say, "Come in," otherwise your dog might mistake your instruction as a recall and run toward you.) Have the visitor open and close the door a few times before entering. Then they should stand perfectly still with arms folded in front of their chest, while you pay full attention to your dog with your back to the door. Keep your dog in a Down on the bed for ten seconds or so, and then instruct, "Say Hello." This cues the visitor to call the dog, "Rover, Come-Sit," while immediately *luring* the dog to sit, taking hold of the dog's collar (to prevent jumping up), and offering lots of *calm* praise and a few pieces of kibble, while also giving a lengthy, *calming* chest rub until the dog is calm and relaxed. The visitor may then let go of the dog's collar and stand up straight while the dog nose-vacuums their clothes. If the dog even looks like they are about to jump up, both you and the visitor instruct together, "Rover, Sit."

Again, it doesn't matter how long this takes or how many times you must reinstruct your dog. Persistently insist, and your dog will learn how to greet visitors in a mannerly fashion.

Phew! That's a lot of attention, effort, and energy on your behalf, but here comes the payoff. Give your dog a few minutes to become bored with the "visitor," and then have the visitor leave by the back door and come around to the front door to practice half a dozen reentries in succession. You will see a dramatic improvement on just the second and third reentries. Remember *Love Actually*. Your dog will be less and less excited with each successive reentry and probably thinking: *Make up your mind. Are you coming or going? I only just greeted you.* By the fourth or fifth reentry, your dog will begin to resemble a mannerly butler and greet your "visitor" in the agreed-upon fashion.

And wonderful for you, with repeated reentries, you'll find yourself praising your mannerly dog more, a lot more, while smiling.

Now it's time for the troubleshooting party. Invite a group of people for brunch and schedule each guest to arrive at different times — ten minutes apart. Instruct each person how to act. Most will be on your team because they have grown tired of being molested by your dog every time they come to visit. Additionally, it really helps if another family member coaches each visitor until they perfect their routine, as you pay attention to and manage your dog.

Each visitor practices six entries in succession before joining the brunch, served by another household member. Once all guests have arrived, one by one, guests may slip out the back door to practice a couple more.

If ten people practiced ten entries, your dog would practice a hundred greetings in just a couple of hours. From now on, if your dog reacts to any visitor, deal with it — instruct, "Bed. Down. Shush." Praise your dog and then ask the visitor to leave and try it again.

After your guests have seen Rover's miraculous transformation over the course of brunch, no doubt you'll be getting a lot of brunch invitations over the next few weeks to transform their dogs' front-door greetings.

On-Leash Walking

Although dogs may *start* heeling or walking by the owner's side, as soon as they move out of position, just a few feet away, the dog's comprehension of the instructions Heel, Sit, or Steady are drastically reduced. An extremely effective technique is to redefine the instruction

Heel to mean "come to Heel position." Then first teach your dog how to walk on-leash at home *before* heading to the sidewalk.

Find Heel

Most people teach Heel when the dog is right by their side, so that's what dogs learn. Additionally, it is important to teach Heel when the dog moves out of position, for example, too far ahead. I use a troubleshooting technique that is very similar to Dawn Jecs's impressively wonderful "Choose to Heel" approach. Practice indoors and in your yard, and it will improve on-leash walking no end.

First, practice the repetitive one-step Heel-Sit routine with your dog off-leash. Now let's add a couple of twists. With your dog on your left-hand side, each time you take a single step, move in a different direction: one step forward, one step backward, or one step to the right. Your dog is now no longer in heel position, so say, "Heel," and lure your dog into a Sit at heel. Then praise and praise some more. You want your dog to learn that sitting at heel is the happy praise place!

To lure your dog forward or to move in closer is easy. What needs practice is when you take one step backward — now your dog is too far ahead. Does that sound familiar when walking on-leash?

First, teach "Back up" when your dog is facing you by moving a lure toward your dog, under their chin, and toward the chest. Then yo-yo the commands "Come fore" and "Back up." Now try it with your dog in heel position. Take one step backward, say, "Back," and then lure your dog to back up to Sit at heel. Keep doing this until your dog jumps back into a Sit. Then alternate one-step forward and one-step backward Heel-Sits to troubleshoot Back up at heel. Both Steady and Back up are lifesavers for leash walking.

Continue to practice numerous one-step-in-any-direction Heel-Sits. Also, maybe try yo-yoing regular heeling and heeling backward. It's insanely funny to watch. Then add a second twist.

After taking one step forward, backward, or to the right, immediately turn 90 degrees, 180 degrees, or 270 degrees, clockwise or counterclockwise, and then say, "Heel." Now your dog is hopelessly out of position and needs to think to *find* heel position, so give your dog a second, and then lure your dog to Sit at heel. Practice this routine over and over indoors and in your yard. Investing half an hour and your dog's entire dinner allotment of kibble will be one of the best investments you make.

Once your dog understands that Heel means *find heel position and sit*, practice with your dog at greater and greater distances. While your dog is wandering around and exploring your house or yard, say, "Heel," and start walking away from your dog. The first few times you may have to repeat the instruction and jolly your dog to entice the dog to follow. As your dog approaches closer to your retreating form from behind, look toward the left and pat your left thigh, praise as soon as your dog comes into heel position, then stand still and praise and reward your dog when they sit.

These exercises improve your dog's comprehension of Heel to mean what most people actually mean, that is, *get back into* heel position, or *Come* and Heel.

First Train Off-Leash at Home

Most untrained dogs learn to pull on-leash after just a couple of walks. A much better idea is to first teach your dog off-leash *at home*, *indoors before outdoors*, and then put a leash on your newly trained dog and go for a walk. If you have a puppy, an adolescent, or adult dog that already pulls, I would just start from the beginning. There are many exercises that improve leash walking. I use a four-step process: (1) off-leash following, (2) off-leash heeling, (3) on-leash heeling, and finally, (4) on-leash walking. I suggest you use your dog's entire dinner allotment of kibble to troubleshoot.

First practice *off-leash following* exercises indoors, from room to room and around the furniture and then repeat outdoors in any fenced area. Developing a psychological bungee cord is the foundation for all leash work; it is vital to prove that your dog *wants* to walk by your side. Start by using toys as lures and lots of food rewards for frequent Sits. Then, be the lure *yourself*. Be animated and use your voice! In a large area, the basic rule for teaching your dog to follow is to *keep moving away from your dog*. If your dog improvises, quickly do the opposite. Change speed and direction often and reward your dog for most Sits.

Then, Heel your dog *off-leash* both indoors and outdoors. When you approach heeling as short Sit-Heel-Sit sequences, it is easier to teach and gives you more control than letting your dog range at the end of a leash. Again, change pace and direction often and have oodles and oodles of Sits with lots of food rewards.

Then, affix a leash to your dog, who now has a reasonable idea of what heeling means. The quicker and more purposefully you walk, the easier it is. Try some longer Sit-Heel-Sit sequences, back and forth and around the perimeter of your yard. Introduce a few "Go sniffs." Then say, "Let's go," and *walk* at a leisurely pace, but give lots of encouragement and instructions, such as Steady, Hustle, Off, and so on. If you lose attention and control, stand still and say, "Heel," for your dog to (come and) Sit at heel. Alternate heeling and walking on-leash, as you will on the sidewalk.

Now go for the first walk in public, but this is not to get from point A to B; it's a training walk. After leaving the house, immediately ask your dog to Sit-Stay-Watch. Then one step at a time, Heel-Sit down your driveway, turn around, and Heel-Sit back to the house. Don't give your dog a second to get ahead of you and lunge. Go back and forth up and down the driveway.

Then move to the sidewalk and stand still with your dog in a Sit-Stay-Watch. It's vital to check that your dog is calm and attentive before walking. Then walk back and forth on the sidewalk in front of your house, so that your dog is always covering familiar ground (rather than continually encountering novel and exciting stimuli with every step). Each time you about-turn, go a little farther. As the back-and-forth distances increase, try longer Sit-Heel-Sits, alternate heeling and walking, and incorporate sniff breaks. As back-and-forth gets longer and longer, eventually, you'll walk all the way around the block. Happy heeling and wonderful walking!

Greeting Others on Walks

As above, the best way to practice passing and/or greeting people on walks is through a dedicated troubleshooting session. Enlist family, friends, and neighbors to participate. The more the merrier. Have the people space out and stand still on a long section of the sidewalk in front of your house. Initially, Heel-Sit back and forth past the stationary people with no greetings. Then, stop by each person to shake hands while your dog remains in a Sit-Stay. Then have the people mill around or walk back and forth, while you do the same but each time a person approaches, or overtakes, instruct your dog to Sit-Stay-Watch.

Next up is to teach your dog that the *default greeting* with any unfamiliar dog is to Heel by, or to Sit-Stay-Watch to allow them to pass by with *no greeting*.

If you asked me to pick just one exercise that dog owners are begging for in every training class, I would say Concentric Circles — this is the all-time best, on-leash troubleshooting exercise to teach mannerly leash walking and mannerly greetings with strangers and their dogs.

Concentric Circles

This exercise works best with lots of participants. In workshops, I've done it with eighty dogs. I would love to see this as an informal exercise in every dog park, like basketball pickup games, in which people may join while the exercise is in progress. Or every neighborhood could coordinate their own with all local dogs participating. I mean, which dogs do you walk by the most? Neighborhood dogs. Or a group of owners could get together and practice.

Split everyone into two equal groups. If you have, say, twenty participants with their dogs, divide them into two groups of ten and then put ten cones in a big circle. Have two people with their dogs stand by each cone, with one person and dog outside the cones and the other person and dog inside the circle of cones to create two concentric circles. If there is an odd number of people, assign the extra person to the inner circle.

People in the *outer circle* walk *clockwise* on the outside of the cones, whereas people in the *inner circle* walk *counterclockwise* inside the cones. Everyone walks/heels their dogs on the left-hand side, so that the *people* in the two circles pass each other shoulder to shoulder, while their dogs are separated by the two owners.

People walk without stopping. It is crazy at first, but after three or four laps, the dogs learn that the default greeting for approaching owners and dogs is *no greeting*; they just keep walking by. It helps enormously that, as they pass each other, dogs are separated by the owners. Give lots of praise and a food reward for every successful pass-by.

Walk until the dogs get the idea. For some people, the first few laps can be quite an effort, as their dog tries to jump on every oncoming dog and hog-tie their owners with the leash. However, this is much easier than walking down a sidewalk, where approaching dogs come occasionally and unpredictably, the owner is unprepared, and the dog is Champions League–winning crazy with each new encounter. In this exercise, each owner must be vigilant the entire time because other dogs are approaching nonstop. No dog can maintain over-the-top and out-of-control excitement indefinitely. Indeed, with each lap they become less excited and more controllable as they pass by the same people and dogs over and over.

Once the group has a semblance of order, the trainer instructs, "Halt." Then each person in the inner circle instructs their dog to Sit-Stay-Watch. Praise and hold a food reward or mega-secondary toy in front of your face to keep the dog focused. The people in the outer circle keep walking until they are alongside a person in the inner circle (hopefully with their dog in Sit-Stay), whereupon they instruct their dog to Sit-Stay-Watch.

All people stand still and focus their attention on keeping their dogs watching them while in a Sit-Stay. Praise lots and reinstruct if necessary. On subsequent laps, people may introduce themselves and chat, but without taking their eyes off their dogs. A few more laps and people may shake hands, while the dogs remain in Sit-Stays without greeting each other.

After many more laps, *if both owners agree*, the dogs may be allowed to greet each other. Hold the dog's leash a foot from the collar, so that the leashes don't tangle. Also, only let the dogs' greet periodically, so that for the most part, they maintain the default greeting as *no greeting.*

A variation of Concentric Circles is for people in one circle to remain stationary and practice stays, while people in the other circle walk by. After several laps, change roles. The more laps you do, the calmer and more attentive the dogs become. After twenty or thirty laps, sidewalk walking will seem so much easier.

In future classes, you can make the exercise much more challenging by reversing the direction of the circles: The outer circle walks counterclockwise, and the inner circle clockwise, with all dogs heeling on the left-hand side, such that *the dogs* pass shoulder to shoulder and are not separated by their owners.

Concentric Circles can be amusing to watch because *owners* often have great difficulty setting up for the exercise, and so they end up walking all over the place like lost sheep. For the trainer in charge, it's like herding cats. However, despite initial disorganization, this exercise is so incredibly effective because with twenty dogs and people, after thirty laps in twenty-five minutes, each dog has experienced a total of 1,140 total pass-bys! (Each dog passes each of the other 19 dogs twice per lap, for 38 pass-bys per lap times 30). That's what troubleshooting is all about.

Other Troublesome Scenarios

The same troubleshooting process applies to all other situations where our dog's unbridled excitement and enthusiasm destroy our control. No matter what trigger messes up your dog's brain, repeat the incident over and over until the dog's brain floats back to earth. With each repetition, a dog will become calmer and more focused.

One common issue is simply leaving the house with a dog. In this case, leave and enter the front door over and over, integrating many Sits in the process: As you leave, Heel-Sit, Heel-Sit, Heel-Sit up to the door and Sit-Stay-Watch as you open the door. Sit-Heel-Sit through the door and then an immediate Sit-Stay-Watch the moment you're outside, while checking you have your keys before closing the door. To check that you have your dog's full attention, it's always good to have a lengthy Sit-Stay-Watch as soon as you step outside, and another at the boundary of your path/driveway and the sidewalk. Repeat the process to reenter the house, and then repeatedly exit and enter your house several times. Once your dog can leave your house in a mannerly fashion, go for a walk and maybe practice entering and exiting the dog park's outer and inner gates over and over.

Sometimes certain words trigger dogs to erupt, such as *walkies* and *dinner*. Some people try to replace trigger words, while others spell them out. But then border collies learn how to spell, and Malinois autofill after two letters. I like to desensitize dogs to the words themselves: I'll sit in an armchair reading, and without moving or looking at my dog, say, "Dinner, dinner, dinner, dindins, kibble, treats, dinner...." Dogs explode with enthusiasm at first and then eventually settle.

If your dog acts up before or during a meal, use the delinquent waiter routine (pages 121–22), which integrates oodles of Sit-Stays into each multiple-course meal.

Remember, the first time is always the most difficult, so repeat the situation over and over. This doesn't just make your life easier. It teaches dogs to contain their enthusiasm and maintain impulse control, which makes all training easier.

Chapter 12

Building Reliability

To illustrate what dogs do and don't understand, I devised a series of simple tests. The first was the Sit Test. Nothing fancy; the test involves no bizarre or frightening distractions. I chose Sit because it's the easiest response to teach on cue, and it's usually the first command that every dog learns. Moreover, it's the instruction that everyone is convinced that their dog "knows." When a dog doesn't comply with Sit, people can get wildly annoyed, even angry, because they assume the only explanation is that their dog is being willfully noncompliant.

In workshops, I would select the best-trained dogs as contestants by calling out randomized position changes with variable-length stays. It was like a marathon dance contest: If a dog had not responded correctly before I called out the next position change, owner and dog would leave the floor. The last eight dogs would be the contestants. Back then, most were Obedience Trial Champions, and I would let all eight demonstrate their exceptional skills, so everyone knew how well they were trained.

I loved obedience dogs as test subjects because I had designed the Sit Test to be especially challenging for well-trained and seasoned obedience competitors. Many obedience dogs are pattern-trained (the same sequences are repeated over and over), and performance reliability often breaks down when dogs anticipate responses. That's all the test involved, minor variations in what the dog and owner expect.

The purpose of the Sit Test is twofold: First, it illustrates that even the most well-trained dogs don't understand the word *sit* as a child might, and second, it proves without doubt that the most common reason for noncompliance is lack of comprehension of a very simple instruction. I would ask the workshop audience to be the jury and decide if a dog who fails to comply is guilty or not of being dominant, dumb, stubborn, angry, or intentionally dissing their owner.

The Sit Test

The Sit Test is simple: To pass each exercise, the dog has to sit just eight times. On each occasion, the owner may give only a single verbal command or handsignal to Sit, and the dog must sit within two seconds and remain sitting for three seconds. That is it.

Obedience competitors are used to being restricted to a single command. However, in the ring, a two-second Sit would be considered very slow, but I give the extra time for the dog's word processing. If the dog doesn't sit, the owner may repeat the verbal command or handsignal until the dog does.

Most of the sits are simple stuff: sits at a distance and when the trainer or dog is in motion. However, no dog has ever passed the test on the first attempt. Some dogs break stays before the handler can even get in position to say "Sit," many dogs don't sit when requested, and some behave unexpectedly. If a dog responds unreliably in the controlled environment of this test, then they might also respond unreliably in real-life situations. Most of the time, this is not a big deal, but sometimes it is.

For brevity, I am just going to discuss three exercises: the Sit-on-Hand, the Down-Stay Sit, and the Back-Turned Sit. Whereas the first five tests simply expose pattern training and cause dogs to incorrectly anticipate what comes next, the dog's responses to these three tests are fascinating and reveal other possible underlying issues in comprehension.

- **Sit-on-Hand:** The handler instructs the dog to Stand-Stay, walks about eight feet behind the dog, lies down in a supine position, with one arm extended with the hand (palm upward) on the ground between the dog's hind legs. The handler instructs the dog, "Rover, Sit."

- **Down-Stay Sit:** The handler instructs the dog to Down-Stay, walks six feet in front of the dog, and lies down on their back with arms crossed over their chest. Thus, both dog and handler are lying in a straight line, with the dog prone and the handler supine, with the top of their head just two inches from the dog's nose. The handler instructs the dog, "Rover, Sit."
- **Back-Turned Sit:** The handler instructs the dog to Stand-Stay, walks six feet away, turns their back to the dog, and closes their eyes. The judge holds the dog on a loose leash. The handler instructs the dog, "Rover, Sit." When the dog sits, the judge praises the dog, "Good dog, Rover," to let the handler know the dog sat successfully. Then the handler may turn around to praise and pet their dog. If the judge doesn't praise, the dog didn't sit.

In the Sit-on-Hand, very few dogs sit on their owner's palm. When the owner gets on their hands and knees to lie down behind their dog, many dogs break their Stand-Stay and swivel around to see what the owner is about to do to their rear end. Other dogs don't sit when instructed. If this happens with the dog's owner, we would expect it to happen with a stranger — for instance, if a veterinarian knelt on the ground to examine the rear of a giant breed to avoid lifting the dog onto an examination table, or if a child crawled behind a dog in any stay. It's important to proof all stays in all sorts of settings and with all sorts of people acting in every conceivable manner.

In the Down-Stay-Sit, when the owner is on their back in front of the dog, not only do many dogs not sit, but they also display all sorts of alternative behaviors. Some dogs crawl over and lay down beside or on top of their person. Others jump up and explode into a wonderfully creative, improvisational flurry of staccato behaviors, such as circling, spinning, wiggling, pouncing, prancing, and playbowing, only to land on the person's chest and lick their eyeballs as a finale.

As I've explained, dogs attend preferentially to body language, and a down-on-the-floor owner is usually interpreted to mean either sleepy time or playtime. Many dogs break their Down-Stay the moment their owner starts to lie on the floor and simply don't "hear" the verbal instruction. So unless we proof Down-Stays with people in every conceivable body position, or displaying any and every activity, we can expect a similar response from our dog when we lie on our back in the sunshine during a picnic, or when there are lots of children rolling on the floor.

In the Back-Turned Sit, when the owner is six feet away with their eyes closed and back turned, very few dogs sit on the first command. Some dogs require three or four commands, and others don't sit after five commands. At that point, I would instruct the owner to open their eyes, turn around, and instruct Sit once more. Invariably, all dogs sit immediately.

The Jury Weighs In

If five verbal requests are given with no response when the owner's back is turned, but the dog immediately sits once they can see the owner's front and face, then this is not a disobedient dog. In every workshop, nearly every onlooker agrees, and it is the dog's behavior when not responding that is most convincing for the jury.

I remember one German shepherd who began to look somewhat perplexed when his owner stepped away and turned her back. Then when she spoke, his massive, radar ears erected as he intently stared at her butt. After the second instruction, his ears seemed to get even larger, and he cocked his head from one side to the other. After the third instruction, he pulled his head back, still riveted on her butt, and slowly opened and closed his muzzle, as if he were mouthing the words, *OMG — it spoke!* Too hilarious for words. (We always used to video these exercises to play back to the owners afterward.)

This was not a distracted dog. The dog was intensely focused on his owner's voice (and her butt) for the duration. He made little intention-twitches and butt-bobbles whenever she spoke, as if he knew he should be doing something, but had no idea what. But he *didn't sit* until she turned around, whereupon the handsome boy responded instantly to her first request. The dog's entire performance was magnificent. As defense attorney for the dog, I interpreted what we had all seen, and the workshop jury's verdict was unanimous:

First-Degree Flagrant Disobedience: Not guilty
Second-Degree Dominance, Stubbornness, Spite, or Daydreaming: Not guilty
Misdemeanor Dissing His Owner in Public: Not guilty

The Sit Test is important because it demonstrates that even exceedingly well-trained dogs don't always sit when their person's back is turned, and yet the person is only two yards away. If that's so, how can we expect compliance when our *dog's* back is turned, and our dog is *forty yards away* and *chasing a rabbit*? How can we expect compliance when calling our dog when out of sight? There are a gazillion scenarios when we are not standing toe-to-toe and face-to-face when we give an instruction.

Building reliability is simple: First, we teach dogs the meaning of our words, and then we proof those commands until our dog knows they apply in every scenario — most especially, when not face-to-face, when we are not close, and when dogs are not already focused on us.

After each Sit Test, we trained and proofed all the dogs to perform each exercise, so they could retake the test and achieve a perfect score. The test locates potential problems and resolves them. In real life, though, there are many more "tests" than these eight. That's why, throughout this book, I suggest that you integrate training into everything you and your dog do together and practice everywhere you go. Repetition in a wide variety of situations is what "proofs" commands so dogs respond reliably most of the time, even when the unexpected pops up. After all, we can't anticipate *every* situation.

The Message of Our Manner

Something else is going on during the Sit Test. Having each contestant/trainer stand in the middle of the room with a hundred observers looking on can melt their confidence, which corrupts their body language and tone of voice. Few contestants give confident instructions for all eight exercises, especially when instructing Sit with their eyes closed. Instead, some ask pretty-please for their dog to sit, and others shout. A person's uncertain manner has a huge bearing on their dog's performance (or nonperformance). Dogs know whether we "mean it" or not.

For example, in a puppy class once, I asked a young fellow to call his puppy from play. He slowly uncrossed his legs, put down his phone and water bottle, stood up, and called, "Rover? Rover? Rover? . . ." The puppy ignored his owner and continued playing. Was the puppy being disobedient? No. The owner had no confidence, and his words were more of an interrogative than an instruction. In fact, he never gave an instruction.

I told him, "Look, Rover knows his name, which is defined in your Doggy Dictionary as, 'Rover, follow the next instruction.' If you keep repeating his name, you're saying, 'Follow the next instruction . . . the next instruction . . . the next instruction . . .' You sound like a broken record. Instead, tell him *what* you would like him to do."

Now the fellow said, "Rover, Come!" But Rover still ignored him. Again, was this the dog's fault? No. The young man hadn't convinced the dog that his instruction was worth paying attention to, that the instruction was even remotely relevant for the dog. I suggested, "If he doesn't spring into action after you ask him to come, do something to get his attention. Squat down with open arms, clap your hands, slap your palms on the floor, squeak, jump like a frog, anything. But do something."

After the fellow called once more, he jumped like a frog, and a bunch of puppies, including Rover, bounded into his open arms.

Until that moment, Rover hadn't learned that there was any reason to respond when his owner said his name followed by meaningless words. *The owner hadn't done anything to make himself interesting to his puppy, and he never bothered to follow up when the puppy failed to respond.* In a sense, the owner was conditioning his pup *not* to respond.

Once the fellow made himself interesting, by acting like a frog,

Rover responded. However, that's not how anyone wants to behave to get their dog to come. Ultimately, we want our body language and tone of voice to be irrelevant, or more specifically, we want dogs to respond to what we say.

Let's go back to what the man did prior to his first attempt to call his pup — he slowly uncrossed his legs, put down his phone and water bottle, stood up, and then called. These actions were all normal *intention signals* indicating that he might follow up. (Although he didn't.)

Surely, we don't want to put down our drink and stand up every time we call a puppy. So I asked him to sit down, cross his legs, hold his capped water bottle in one hand and another unbreakable object in the other, and to look relaxed and preoccupied, and then do the following:

1. Calmly say, "Rover, Come," and wait half a second.
2. Then, if Rover didn't come, he was to drop everything, abruptly stand up, and perform the effective squeaky-frog routine.

The man did it, and Rover came within a couple of seconds, along with a couple of other puppies. All the pups then received bountiful praise and hugs before being instructed, "Go play."

Dogs learn all the clues that signal when they do or don't need to comply — the times, places, situations, and most importantly, the tone of voice and body language that advertise whether *we are likely to respond* and follow up if they don't. What are some of those clues? When we're sitting or lying down; when we're cooking, eating, drinking, holding a baby, texting, reading, watching television, making out, in bed, or making out in bed (to make another baby to hold). There is a veritable bounty of signals that tell dogs, even if they understand the words we say, those words aren't relevant to them in that moment.

The follow-up squeaky-frog routine is a way to teach dogs that what we say is relevant to them. After just a couple of repetitions, you'll never have to act like a frog again because after you calmly say, "Rover, Come," your dog will bound toward you, and so there will be no need to follow up. Praise and say, "Go play" instead.

Please don't feel upset if your dog ignores you. Most of the time, this is a subconscious process, a way of filtering out the sheer volume of words people say to dogs, to keep them from cluttering up the dog's consciousness. There's nothing wrong with your dog's hearing, and we can test to confirm comprehension. It's just that words classified as "irrelevant" never reach the conscious levels of a dog's brain. Dogs are not intentionally making fools out of us. It's just that most of what we say *is* irrelevant to dogs, and by not following up, we further confirm that.

Yes, at times, a dog might purposefully ignore a request, but this is usually when the dog is comfortable or preoccupied with some higher mission — such as playing, wandering, watching, exploring, sniffing grass, sniffing another dog's butt....

We also need to acknowledge that life is a two-way street. Lando, one of my favorite dogs in training, relished long relaxing evenings on the couch. He was immediately responsive to *some* of my suggestions, such as "Wanna bit of popcorn?" or "Fancy a chest scratch?" But when it came to a final pee and poop before bedtime, he was completely and utterly unresponsive to anything I said or did. He didn't even squeeze his eyelids tighter to make me disappear. I *could* get him off the couch and moving toward the door with my voice, but knowing that I could, I no longer did. I didn't want to rudely crash his sleepy brain with an intrusive squeaky-frog routine at 11 p.m.

Instead, I would gently move his wheat-sack body closer and parallel to the edge of the couch, so his legs stuck out into space. Being careful not to bend him in the wrong places, I carefully rolled his body and placed each foot on the ground, at which point his four-on-the-floor supported his own weight and he activated, took two steps, stretched fore and aft, and then trotted merrily off to pee and sometimes poop. Afterward, he joined the other dogs sitting in the treat jar circle, and then he happily trotted off to his own bed and fell asleep within seconds. *Of course*, he was training me, but I loved nursing him up; that was one of our rituals.

Most trainers insist that we must be consistent, as in consistently consistent all the time. But that's unrealistic. I don't know too many people, including myself, who can maintain consistency for lengthy periods, let alone 24/7. Being consistent for lengthy periods is extremely taxing, and inconsistencies are unavoidable.

Also, we don't want our dogs to be on edge, vigilantly alert, and at our beck and call twenty-four hours a day. We want our dogs to be able to relax most of the time yet be immediately responsive to our requests when it's important to us or an emergency.

So how do we clearly communicate to our dogs those instances when they must respond immediately and reliably?

The Dog-Con System: Three Names for Your Dog

When Omaha was just six months and four days old, he was in an obedience ring for his third time during a three-day dog show weekend. Our first two trials were a breeze, but the third was a total bust. During the off-leash heeling exercise, he nonchalantly wandered off and hopped over the ring fence, then over three more fences, and made a sauntering beeline to visit another competitor. He joined a golden retriever during off-leash heeling exercise, walking between the owner and the frisky female, alternating his gaze from dog to human. Humiliation! (So lucky I seldom get embarrassed.) I was left in the ring still walking the prescribed heeling pattern on my own, pondering: *If only I could explain to Omaha the full import of our brief ten minutes in the ring!*

I mulled over the problem as we chatted afterward, both of us lying down in the grass. He really seemed interested in what I had to say, although I'm certain he didn't understand more than a few words.

Later that evening I read some poetry, and a delightful poem by T. S. Eliot entitled "The Naming of Cats" included a line that caught my eye: "a cat must have THREE DIFFERENT NAMES." That was the solution! Omaha needed three different names.

Usually, we use our dog's name to indicate who we are talking to. But if Omaha had three names, I could use them *to code instructions in terms of urgency and importance*. A lot of parents do this. At a toy store, I once heard a woman speaking to her son: "Johnny, don't touch. Johnny, careful. Johnny. Put. That. Down. John-neee, you'll break it. Johnny... Juan-Carlos Hernández, *sientate*!" Little Johnny sat!

At home I usually called Omaha "Ohm," a single unit of resistance, which in many ways seemed appropriate for a young male Malamute. His AKC name was Totemtok's Omaha Beagle, or plain Omaha for short. So what I needed was a third, showtime name for competition. I chose Wahoo. Now, all I had to do was teach him the meaning of his three names. I dubbed this the Dog-Con system and then set about teaching Omaha the differences between his three names, Ohm, Omaha, and Wahoo.

Dog-Con 3: This is a dog's *informal nickname*, and using it signals to the dog: Here's a suggestion, and it would indeed be wonderful if you followed the suggestion, but compliance is *not* necessary. For example, if I said, "Ohm, come'ere," or "Ohm, settle down," and he didn't come or settle down, that was cool because he knew that my instruction was just a suggestion, and that I was *not* going to follow up. In a sense, I *formalized and allowed inconsistency* — for both of us.

Dog-Con 2: This is a dog's *formal name*, and it's meant to be used for immediate on-demand compliance in daily living. Saying this name communicates that the next word is an instruction that must be followed, immediately, and with no exception and no discussion. For example, "Omaha, Sit" meant stop whatever you are doing and immediately Sit-Stay-Watch. Further, I was signaling that I was *paying attention*, and that *I would always follow up*.

Dog-Con 1: This as a dog's *competition name*, one that's reserved for performance. When I used "Wahoo," I was saying: *Omaha, Mr. Reliability, it's showtime! Give it your best shot, with perky ears, an eager trot, and a flourishing tail. Watch me as appropriate, and please do look like you respect and admire me. Pizzazz and precision required. I'm trying to give my best performance, too.*

Using the Dog-Con System

Most of the time, I used Ohm, which allowed both of us to relax and be ourselves for most of the day. Occasionally, when visitors arrived, or on walks and in dog parks, I would use Omaha when a higher level of control was required around people and other dogs. I also use a dog's Dog-Con 2 name when proofing reliability and teaching distance and emergency commands.

Meanwhile, Omaha's Dog-Con 1 name, Wahoo, was reserved for training for the show ring, competing, television interviews, and demonstrations when reliability, precision, style, and sparkle were required.

If your dog doesn't need a "performance" name, then just use a Dog-Con 3 nickname and a Dog-Con 2 formal name. However, do consider competing in AKC Obedience at the Novice level, Rally, or Agility trials, since *your dog will love them all, and so will you.*

However, when teaching reliability, I don't want the dog's formal, Dog-Con 2 name to be like "Juan-Carlos Hernández" — the last-straw, Mom-bringing-the-hammer-down name. This is not a "punishment-is-coming-if-you-don't-sit" name. Juan's mother was angry. The purpose of a formal name is not to threaten or express frustration, but to simply communicate that *compliance is required and we will follow up.*

Some working dog and obedience competitors use a similar technique by speaking in English with their dogs at home but using French or German when competing, which is also extremely effective, and it prepares you for holidays in France and Germany. *C'est wunderbar, si?* Other trainers use a competition name and different competition commands.

The Dog-Con system uses different names to code our instructions, rather than using different languages or different words for instructions. By simply transitioning back and forth between "Ohm" and "Omaha," I was able to allow both of us to relax most of the time and embrace our inconsistencies, while preventing our daily inconsistencies from trashing training reliability. Having at least two names for your dog effectively preserves on-demand, ultra-reliability when needed.

It is confusing and unfair to allow dogs to ignore us most of the time, only to get on their cases at other times when dogs don't respond. It is so important for owners to clearly communicate to their dog *when* compliance is required and when it's not.

When you use a nickname, your dog knows it's just a suggestion that, for their own mysterious reasons, they may choose to follow or not follow (much like a stop sign in Paris). There's no reason to ruin your dog's day or spoil a luxuriating sniff if you don't have a good reason. However, if you use your dog's formal name, your dog knows that you expect them to stop sniffing and come immediately, if only for your own mysterious reasons, and that you will follow up.

I think the principle underlying the Dog-Con system offers a more realistic and achievable solution to the consistency conundrum: Relax and be as inconsistent or consistent as you like much of the time, but consistently pay attention and always follow up after giving any formal instruction.

Teaching a Formal Name

Teaching your dog they have an *informal* name is easy. Carry on as normal. Continue to use whatever name or names you've been using most of the time, and anoint them as nicknames, for example, Brandy or Snifter. Then decide on a *formal* name. Maybe their full given name — Hennessy Hound. Or maybe choose a new name — Fido. The names are your choice.

To introduce a *formal* name, first check your dog's comprehension of basic instructions, and then, out of the blue, say, "Fido, Sit," and if Fido doesn't sit, *follow up*. There's no need to change your tone of voice or manner. There's no need to get gruff or shout. A formal command is not a threat of disciplinary measures. Rather, it signals to the dog that you expect compliance and that you will *follow up*.

Very soon a dog learns the difference between their nickname, "Rover" — which signals off-duty suggestions — and their formal name, "Fido" — which signals on-duty, formal commands. Indeed, dogs learn to act differently whenever we use their formal name because *we* act differently when we use their formal name — we follow up. Just as we are capable of being relaxed and distracted much of the time, yet snap to attention when required, so can our dog.

Repetitive Reinstruction Until Compliance

On-leash walks are a great way to practice using both names: As you're walking, teach your dog that "Fido" is their formal name for mandatory Sit-Stay-Watches, which will be the foundation for emergency commands, and to Heel, when crossing streets for example. You can also practice with your dog off-leash indoors. However, to practice off-leash in dog parks, see the instructions in chapter 13 for teaching distance commands.

To teach your dog their formal name, use the informal name while walking — "Rover, Let's go" — and then use the formal name to instruct, "Fido, Sit," or "Fido, Heel." If your dog sits immediately, following a single formal command, praise and pet them in a heartfelt way, and say, "Rover, Let's go," and the walk continues. However, if your dog does not immediately sit following a single formal command, use this simple two-step process:

1. Repeat the formal command using the formal name, "Fido, Sit," as many times as necessary until eventual compliance, that is, *persistently insist* until your dog eventually sits.
2. Once your dog *eventually* sits, pause for a few seconds, offer a terse "thank you," say, "Rover, Let's go," and take one step forward, and as your dog starts to move, immediately repeat, "Fido, Sit." Dogs tend to comply quickly because, after the previous Sit, you have your dog's attention. Praise sincerely and say, "Rover, Let's go," and the walk continues.

However, if your dog still doesn't comply immediately following a single formal command, repeat steps 1 and 2 as many times as necessary until your dog does. Then, once your dog *responds immediately following a single formal command*, praise and pet in a heartfelt way, say, "Rover, Let's go," and the walk continues.

Just to clarify, while I use different commands with the different names ("Rover, Let's go," and "Fido, Heel"), you may use either formal or informal names with most common instructions, like Come, Sit, Down, Stand, Heel, Let's go, Shush, Settle, and so on. That said, when walking, I find myself almost invariably reserving Heel as a formal instruction and Let's go as an informal instruction.

Basically, giving a formal command to Sit presents a dog with a choice:

1. To sit immediately following the single formal command, and so receive profuse praise, and the walk resumes immediately.
2. To not sit immediately and so have to Sit a *minimum of twice* before the walk resumes. That is, by not complying immediately the first time, the dog has to repeat Sit *at least* once more, but perhaps several times, until they eventually Sit immediately following a single formal command.

Dogs learn that, when a formal command is given, all other activities will be put on hold until the dog complies the way we want. Dogs learn that the best way to avoid extended interruptions to ongoing activities is by immediately complying with every formal instruction, which is signaled by their formal name. Consequently, response-reliability percentages for formal commands markedly improve.

My grandfather once used this exact same technique to adjust my lack of manners. Grannie had cooked Sunday lunch, and Gramp was about to carve the pheasant. He said, "Ian, sit down, please." I know I heard his words somewhere in the depths of my brain, but I was preoccupied and didn't sit. After a while, I looked up to see that Gramp had put down the carving knife and fork and was looking at me. He said, "Ian, sit down, please." I did so, but then Gramp said, "Ian, stand up, please." This seemed weird, but I did. Then Gramp said, "Ian, sit down, please." *Got it.* I sat, and Gramp said, "Thank you, Ian," and resumed carving.

Fears of Learned Irrelevance

The mere notion of repeating a command has long been an utter and absolute no-no in the dog training world. Trainers often tell people, "*Never* repeat a command!" And when I suggest *repetitively* repeating instructions, trainers will scream, "Nooooooooooooo!" even though the process is an inherent aspect of building reliability and increasing the relevance of formal commands.

Why do so many dog trainers object to repeating instructions? There are several reasons. They say, if we repeat instructions, the dog will wait for multiple commands before obeying; that training will be more difficult and take longer; that it's nagging; and that dogs will learn that instructions are irrelevant.

First, the notion that "repetitive reinstruction until compliance" causes dogs to wait for more instructions is not true. Observation and quantification prove otherwise. We find that with progressive trials, the required number of commands steadily *decreases* until eventually dogs respond after a *single formal command* practically all the time. Moreover, from the very first attempt, dogs respond following a single command on *every* trial. Remember, if they don't do so after the initial command, we keep repeating the exercise until they do.

Additionally, repeating a cue over and over with no consequences does not make training more difficult or take longer. I have proved this many times in workshops. I would pick a new cue, such as "Beg" and repeat the words twenty times without requiring anything of the dog. Then using lure-reward training, I'd teach the dog to sit up and beg using the same verbal cue. The dogs who had heard the cue many times before training learned just as quickly as dogs that had not. In essence, *all human words are irrelevant* to dogs until we make them relevant. We cannot make irrelevant words *more irrelevant*. Training is all about taking irrelevant signals — words, handsignals, clicks — and making them matter to dogs.

Nagging is a completely different kettle of fish — that is what I'd call "repetitive reinstruction *without* compliance." Nagging creates noncompliance, whereas, by definition, "repetitive reinstruction *until compliance*" builds compliance.

Learned irrelevance applies to a *once-relevant cue* that becomes irrelevant. For example, if we taught a dog to Sit on cue, but then through nagging (repeating the command without enforcing compliance), the dog *learns* that the cue has now become irrelevant. Perhaps the most common dog training example of learned irrelevance occurs when people send their dog away to be trained. The dog learns to respond as required with the trainer, but after just a few days at home, the dog quickly learns that their owners do not follow up, and so relevant commands with the trainer in the kennel become irrelevant in the dog's household — *situational* learned irrelevance.

Repetitive reinstruction until compliance is the exact opposite. Before the procedure, *all* verbal cues have some measurable response-reliability percentage. Repetitive reinstruction until compliance *increases* the response-reliability percentage of these cues when they are used as formal commands.

In fact, repetitive reinstruction until compliance is "learned *relevance*." Obviously, the ultimate goal for any formal command is *instant compliance following a single instruction*. However, I think many people confuse the process of training with the final performance. Training is an educational process, and repeating the cue is a teaching tool to achieve the end product. Tracking the increase in response-reliability percentages and the expected, concomitant decrease in the number of required cues proves this, as do real-life situations.

Real Life

With young pups on their first night at home, I have sometimes repeated the words *settle down* and *shush* fifty or a hundred times. The pup doesn't even understand the words yet. However, I hold on to a piece of kibble and use its smell to try to lure Settle and Shush, and then I softly praise, sing, or recite nonsense poetry to further soothe and mesmerize the pup as they grow calmer and quieter. As I seduce the pup to settle down and sleep many times over, gradually the pup begins to get meaning of the words. Sometimes, it takes hours, but the following night, the pup requires less and less time and fewer and fewer instructions to stop vocalizing, lie down, and fall asleep, and I am soon back to sleeping in my own bed again.

What about real-life emergencies? Of course, we repeat instructions. Let's say we are a passenger in a car and realize the driver is about to run a red light. We don't say "stop" once and start praying for fear of creating learned irrelevance. We say, "Stop! *Stop! STOPPP!!!*"

Or, if our dog is trashing a children's birthday picnic across a field, we don't shout, "Rover, Sit!" once, and if our dog doesn't comply, pretend we don't know them. Of course, we repeat the command; we always do in real emergencies, and we often do during training.

Consider how a copilot checks the status of a plane during landing. They ask questions and give instruction in a calm voice and repeat and add emphasis as necessary: "Altitude? Bearing? Altitude? Flaps? Landing gear? Lann-ding gear!!" Even the on-board computer does the same: "Landing gear. *Landing gear! Altitude! PULL UP! PULL UP! PULL UP!!!*"

Both pilot and plane comply and pull up to fly another day.

Chapter 13

Teaching Distance Commands

Teaching and testing the response-reliability percentages of formal commands indoors and outdoors at home, and during on-leash walks, are the foundation for teaching the most useful commands of all — emergency, off-leash distance commands. Once you've taught your dog to Sit or Down reliably at a distance, life changes. Your dog is now much safer when off-leash, and training becomes next to effortless. It doesn't matter how far away your dog is, what they are doing, or what else is going on, you can stop your dog from getting themselves into trouble by instructing Sit, and your dog sits. Then if everything is OK, you can immediately and massively *reward your dog at a distance* by simply saying, "Free dog."

I much prefer Distance Sits and Distance Downs to recalls, particularly in urgent or emergency situations. Your dog's Distance Sit or Down offers instant proof of compliance while you assess the situation or go to retrieve your dog. With an emergency recall, the dog is running through the environment, and a lot can happen on the way. You won't be in total control until your dog gets to you. Moreover, Sit and Down are easier to teach and easier for dogs, requiring less effort than running all the way back to you.

When my dogs are off-leash, I supervise them all the time for safety, but I also use off-leash exploration and play as an ongoing opportunity to frequently reinforce distance commands. Every couple of minutes, I ask the dog to Sit, and a good 95 percent of the time I immediately release them to resume their interests. The sheer number of life rewards out there is astronomical, and we can use them all to reinforce dogs at a distance. My goal is to make Distant Sits by far the most reliable response in a dog's behavioral repertoire, so I can use it as my go-to emergency command.

But also, I love watching all dogs play off-leash. It's an unparalleled chance for them to enjoy being dogs.

Why Not Use an Emergency Recall?

When I trained Omaha, I used a recall as an emergency command, which worked well because he was a slow-moving dog. Phoenix, however, was fast. I trained her to be the fastest Malamute on planet Earth — in part because I was reacting to jibes about Omaha's reliable yet measured approach in obedience.

So far, so good. Then one day when Phoenix and Oso were playing in Codornices Park, I called them, "Dogs, Come." ("Dogs" was their combined formal name.) Phoenix was streaking toward me like a wolfy missile when I realized a group of elderly people heading toward a children's softball game had moved within her trajectory. Several parents screamed, and I thought it was going to be skittles. I screamed, "Dogs, *DOWN!*" I had never used an emergency Down before, but both dogs hit the grass. I ran up to my dogs, apologized to everyone present, and then asked Phoenix and Oso to Sit, Bow, Beg, and High five to show people that they were friendly. Then, the softball game was put on hold while all the children, one by one, came to say hello to Phoenix and "Handsome." Then all the elderly folk said hello, and then some of the parents.

The incident convinced me that a recall is not always the best choice for an emergency command. In an emergency, it's unwise for a dog to come flying through the environment. An emergency Sit or Down is a much better choice and gives us more options. Also, in a true emergency, we're scared. We shout, our voice is full of fear, and our volume and tone naturally inhibit movement. Maybe with my next dog, I'll use "Bang!" for an emergency command.

I prefer teaching an emergency Sit rather than Down for a few reasons: Sitting is usually easier and quicker for most breeds, especially larger breeds; people can see a dog better when sitting rather than when flat on the ground; and a sitting dog looks less menacing to onlookers. That said, an emergency Down works just as well. It's your choice. I also know some mastiff and sight-hound owners who use "Whoa" (Stand-Stay) as an emergency command. The main thing is to stop the dog from moving and keep them still.

Once your dog is steady in a Sit-Stay, you have several options:

- If everything seems OK, release your dog by saying, "Free dog." Get in the habit of frequently instructing Distance Sits, so you can frequently say, "Free dog," and so repeatedly and heavily reinforce Distance Sits from a distance.
- If a disturbance is far away, or headed in another direction, you can wait a while to confirm it's safe and maybe instruct Down for greater stability and comfort.
- If the disturbance is heading toward your dog, and the coast is clear between you and your dog, recall your dog. A reliable recall is usually a given when your dog is in a stay and looking at you.
- If a potential emergency is rapidly approaching, or the dog is already in the middle of things — for example, running children or people on horseback — then *you perform the recall*: Run to your dog and put them on-leash as quickly as possible.

Teaching a Distance Sit

The notions of "repetitive reinstruction until compliance" and "persistent insistence" for progressively teaching *learned relevance* lay the groundwork for teaching distance commands. In a sense, when teaching distance commands, we are actually teaching the dog the meaning of brand-new commands. Although we're using the same instruction,

Sit, the dog simply has no comprehension of this word when given out of the dog's proximal comfort zone. This is evidenced by testing response-reliability percentages at increasing distances.

The highest response-reliability percentage for Sit is when both owner and dog are close and facing each other. The second-highest score is usually Sit when in heel position. This shouldn't be surprising, since nearly all Sits are practiced when dogs are in front or heel positions and right next to us. However, *response reliability drops off the farther our dog is away from us*, precipitously for some dogs, and especially when the dog is not facing the owner. That is, the dog's comprehension of Sit reduces with increasing distance until eventually bottoming out at *zero*. Even three feet away, response reliability takes a hit, and for some dogs, it zeroes out at six feet! The distance at which command comprehension reaches zero varies depending on the dog, degree of training, and the circumstances.

Hence, a verbal Distance Sit is essentially a new command for all dogs when past a certain distance. Consequently, we must expand our dog's understanding of Sit so they comprehend that it applies at greater and greater distances, and even when their back is turned. We tackle this in two ways: Start close and move farther and farther away, or start in the zone of zero comprehension and move closer and closer.

Start Close and Gradually Increase Distance

Before you start, check that when standing toe to toe, your dog has response-reliability percentages of at least 90 percent when randomizing Sit-Stay, Down-Stay, and Stand-Stay, and *using verbal requests only*. Then progressively increase your working distance to three feet, six feet, and nine feet away, and so on. Once a dog starts to respond from three or four yards away, we're off and running because a dog is gradually getting the message and will progressively generalize comprehension to greater distances more quickly.

If you've never worked your dog at a distance before, when asked to Sit at a distance for the first time, most dogs will typically run back to you. Obviously, that makes teaching distance commands impossible, especially with fast-moving dogs and/or small dogs that arrive in a flash. This technique works well with larger and/or slower dogs. With faster-moving dogs, it helps to use a small mat or training platform, and if your dog moves out of place, simply instruct, "Mat" or "Platform." Or, to prevent your dog coming back to you, work with your dog in a dog crate, behind a chain-link fence, or tied to a tree.

As when teaching any new request, teaching distance commands follows the same basic training sequence:

1. Distal cue → 2. Proximal cue → 3. Distance Sit → 4. Praise

The new and yet *unknown* distal cue is always followed by the *known* proximal (or closer) cue, which provides the information as to *what* to do. After verbally cuing your dog to Sit from a distance, wait half a second and give a flourishing handsignal to Sit. Even though the goal is for dogs to respond at a distance to *verbal* cues, using handsignals accelerates learning verbal cues.

Start with your dog in a Sit-Stay, step back three feet, then: (1) Say, "Rover, Down," and half a second later give a flourishing Down handsignal, from high in the air and ending as close as possible in front of your dog's toes; (2) wait one second, then step up to your dog and say, "Down"; and (3) when your dog lies down, (4) praise and occasionally reward.

Keep your dog in the Down-Stay while you step back three feet. Then: (1) Say, "Rover, Sit," and half a second later give a flourishing Sit handsignal, from beside your knee to the sky and going up on your toes in the process; (2) wait one second, then step up to your dog and say, "Sit"; and (3) when your dog sits, (4) praise and occasionally reward. You'll likely have to give a much more energetic handsignal to get your dog to sit up from being down.

Then, yo-yo three-foot Distance Sits and Downs. Be on the alert for the first few times your dog responds *after* your distal verbal request and handsignal, but *before* the proximal cue. Each time that happens, praise immediately as you step up to your dog and celebrate, giving several food rewards in succession. Then add Distance Stands into the mix so that you can randomize Sit-, Down-, and Stand-Stays from three feet away.

Once your dog responds to most distal *verbal* cues from three feet away — that is, *after* your distal verbal cue but *before* the distal handsignal — repeat the entire process from six feet away, and then from nine feet away, and so on, until eventually, your dog responds when you are all the way across the room.

When working from greater distances, if your dog does not respond to the initial distal cue or handsignal, quickly walk back to your dog, taking large steps and repeating the verbal cue and handsignal with every step. As you walk back to your dog, with every step you take, your dog has better and better comprehension of the repeated verbal cue and handsignal. Essentially, with every step, you are giving a different and more-easily understood instruction. Never forget, dogs are dogs, and they learn very differently from us.

As training proceeds, your dog will likely respond before you get all the way back to them. Praise profusely and then slowly walk up to your dog to offer a few appreciative strokes and maybe a food reward. As your dog begins to respond from greater and greater distances, on occasion, remember to walk back to stroke your dog and give a food reward; otherwise, your dog may learn that you only come back when they don't respond to your original request from a distance.

After releasing your dog and allowing them to explore the room, continue practicing Distance Sits, now using "Free dog" as a life reward.

Start Farther Away and Move Closer

This variation picks up where the previous one leaves off. It's time to work in a larger area to practice even greater distances. Either use the same procedure as before and randomize Distance Sit-, Down-, and Stand-Stays, or focus on teaching a single, distance emergency command, when your dog is off-leash and focused on their own thing.

I usually go for the gold and practice in a dog park, once a dog has gotten the need for dog-dog play out of their system and is roaming, sniffing, or lounging. I usually give the initial distal instruction just outside of the dog's range of comprehension.

1. Say, "Rover, Sit," followed by a flourishing handsignal.
2. Immediately and quickly stride toward your dog, and with each step, repeat the verbal command and handsignal.
3. At some point, your dog will Sit.
4. Praise, and occasionally reward.

With every step you take toward your dog, they will have better and better comprehension. Moreover, with repeated trials, your dog's comprehension at every step will increase slightly, thus progressively expanding your dog's zone of comprehension.

As proof of your dog's increasing comprehension, each time your dog sits, make note of the distance between you and your dog, and the total number of cues you gave (verbal requests plus handsignals).

It helps to video each training session to score the numbers of cues and the distance. Plot the values on a graph. The first few trials will likely require numerous reinstructions, and your dog probably won't sit until you're very close. But you'll see two distinct trends on your graph: The number of required cues will progressively decrease, and your control distance will progressively increase.

Once your dog is getting the hang of it, give verbal requests only. Eventually, your dog will sit following a single *verbal* request at great distances, and when you continue to practice this regularly, on off-leash trail walks and during every dog park visit, you'll soon have a very reliable emergency Distance Sit.

As I mention earlier, in terms of the *art and science* underlying the basic training sequence, the science is simple and set in stone: New and unknown cues are always followed by known cues, such as lures, handsignals, or in this example, proximal verbal cues.

However, the art of training is creative and always evolving, and it often comprises devising innovative lures. For example, you could use a procedure similar to the one I describe earlier to train without food on your person: Use an assistant and tag-team your dog. Have a family member or friend walk your dog on-leash some distance ahead of you. After you give a clear verbal instruction (and a demonstrative handsignal, if your dog is looking your way), the other person repeats your verbal instruction proximally.

Desensitizing Shouted Commands

When working at extreme distances, it becomes necessary to shout instructions. Similarly, we generally shout (or scream) commands in real-life emergencies. In a genuine emergency, it's hard, if not impossible, to be calm. We shout or scream commands with extreme emotion in our voice. The big question is: How will our dog respond? Will our dog quickly and calmly Sit-Stay as instructed, or will our dog bolt, thinking that we are angry? Consequently, it is essential to teach dogs not to stress when we shout commands at a distance or during emergencies. We must reverse their natural interpretations of tone and volume and teach dogs that a shouted command means *urgency* and not *anger*.

Reverse the Meaning of Tone and Volume

First, practice indoors with your dog off-leash but fairly close. In essence, in a very slow progression, repeat your distance command (either Sit or Down), and in very small increments increase your volume and change your tone. Over the course of thirty to forty gradations, go from a soft, neutral tone to vocalizing a real-life screamed emergency command. For brevity, I've listed only eight progressive steps to give you the overall idea:

1. Say, "Rover, Sit," in a soft voice and neutral tone. When your dog sits, say, "Thank you," also in a soft and neutral tone.
2. Say, "Rover, Sit," slightly louder, and then praise, "Good dog."
3. Say, "Rover, Sit," louder still, then slightly extend the praise, "Gooood dog."
4. Say, "Rover, *Sit*!" in a reasonably loud voice, then praise just as emphatically, "*Good dog!*" and offer one piece of *kibble*.

5. Over a series of increments, eventually shout, "*Rover, SIT!*" and praise just as loudly, "*GOOD dog!*" and offer *three pieces* of kibble.
6. Truly shout, "ROVER, SIT!" and shout your praise, "GOOOOD! DOG!" and offer a *tastier* food reward.
7. Over another series of increments, shouting nearly as loudly as you can, "ROVERR, SIT!!" and pump up the praise, "Brilliant! Marvelous!! Suuuuch a GOOOD DOG!!" and offer five tastier food rewards.
8. Finally, for the last iteration, visualize that your dog is about to run into the street in front of a moving car. Believe this and *scream*: "*ROVERRRR, SIIIIIIITT!!!*" Then praise as if you just saved your dog's life: "*Whatta GOOD BOY! Oh you're the BEST DOG EVER!!! I thought I was going to lose you.*" Then hug, head ruffle, chest rub, and give a veritable smorgasbord of food: freeze-dried liver, chicken, lamb, and so on. Be truly and genuinely happy!

After this routine, your dog will have a different perception of shouted commands: *Wow! I love it when they shout. When I sit, I get MUCH better food rewards and a really happy and appreciative owner!*

Thereafter, return to normal. Issue all formal commands in a soft, controlled, neutral tone, and reinforce your dog's responses with true, meaningful praise and life rewards. Aside from occasional practicing to reinforce this lesson, reserve shouting commands for true emergencies.

Administer Life Rewards from a Distance

Teaching distance commands takes a little time. But once you've taught them, maintaining and further increasing distance reliability becomes effortless and enjoyable because now you have the means of using life rewards *from a distance*. Releasing your dog from a Distance Sit is the most powerful reward you'll ever have under your control, so use it often. Every half minute, instruct your dog to Sit, and then in most cases, release your dog immediately to continue their explorations.

I strongly suggest that you spend half a day in a dog park to thoroughly and absolutely integrate Distance Sits into your dog's off-leash experience. Nearly every training interlude lasts only a couple of seconds, and every reward — "Free dog" — is simply enormous.

This is the stage in training that I love: walking or jogging with Omaha along forest trails, just watching Dune and ZouZou be dogs in off-leash parks, and walking in our "secret place." Every couple of minutes I would instruct Sit to check that I was in control *and to further reinforce reliability*. To be honest, every time I see one of my dogs sit at a distance, I just can't believe it. But they sit so promptly and happily each time. These are special moments, and I love sharing them with my dogs. I feel that I'm on cow-back once more, enjoying the magic of moving Blackie's ears to get her to move forward and change directions.

Part 5

PROBLEMS AND MISCONCEPTIONS IN DOG TRAINING TODAY

Teaching ESL from the outset facilitates all aspects of training: not only for cuing desirable behavior but also for providing verbal guidance to reduce and eliminate undesirable behavior and noncompliance, and hence, the need for aversive punishment.

Aversive techniques have always been the elephant in the room. In chapter 14, in the same way I objectively evaluate the ease, speed, and effectiveness of different reward-training techniques in chapter 3, I do likewise with the use of "aversive punishment." As part of this, I examine the two complicated, confusing, and often misunderstood concepts of "negative punishment" and "negative reinforcement" — all with the express purpose of suggesting easier, quicker, and more-effective, *nonaversive* alternatives.

As always, clarity comes from asking our usual questions: Did it work? How well did it work? How quickly? And what are we trying to teach?

Additionally, misguided breeding practices are a problem that concern us all. Excessive and unilateral selective breeding has harmed the health of purebred dogs, increased the incidence of diseases and conditions, and caused many dogs to die much too young. I highlight these issues and suggest solutions, since there's no earthly reason why dogs shouldn't live to be fifteen to twenty years old.

Chapter 14

Why Punishment Often Fails

Most trainers have strong feelings about aversive punishment. Some trainers insist that aversive techniques offer a quick fix for misbehavior and noncompliance and that they are the only way to deal with extremely difficult and potentially dangerous problems. On the other end of the spectrum, some trainers think that using any form of punishment or aversion is wrong under any circumstances.

As a result, discussions often dissolve into emotional, moral, and ethical arguments about the *nature* of punishment rather than on evaluating its *effectiveness*. This is not to dismiss the issue of ethics in dog training; however, it is much more productive to initially focus on how well, or if, training techniques work. Because if certain techniques are ineffective, discussion is moot. There's no reason for debate. Change to Plan B right away; now renamed as Plan A.

The primary goals of lure-reward training are to teach ESL to (1) increase the frequency and response-reliability of desirable cued behavior and (2) reduce and eliminate undesirable behavior. Many people have a knee-jerk reaction to use aversive techniques to accomplish the latter. However, what *are* the easiest, quickest, and most effective means for dealing with misbehavior and noncompliance?

So much of this book — teaching ESL, cuing and reinforcing desirable responses, and creating nonaversive techniques for preventing and resolving behavior, temperament, and training problems — has come from analyzing all the reasons why physical corrections and aversive punishment so often fail. Once we diagnose the constraints and limitations of aversive punishment, we know what to fix, and almost exclusively, the solution is using our voice.

This chapter reflects my journey — along the ghost trail of some caged rats that were shocked without warning — to come to an understanding of what dog training desperately needs — ESL and heartfelt praise.

Omaha and the Lightbulb Moment

I didn't start to explore the traditional dog training world until I got Omaha, my very own first puppy. Like any parents, Mimi and I started checking out dog schools, and I was shocked to find that the only options were *on-leash* and that Omaha couldn't start *until he was at least six months old!* I panicked. Omaha was already six weeks old, and he would be coming home in two weeks; I needed to plan his education. Eventually, I started SIRIUS Puppy Training so that Omaha could go to school. However, I planned to compete in obedience trials, and so I asked to audit some obedience classes to see what was done. My much-too-young puppy was not allowed to come with me.

I don't know what I expected, but that first obedience class was a shock. I had seen an on-leash class on television, but honestly, I thought the dogs were on-leash to keep them controlled for the camera. I had absolutely no idea that *all* classes were taught that way.

Much of the class comprised repetitive obedience drills. Very little time was spent teaching owners or dogs exactly *what* to do. Instead, owners were instructed to give commands in a clear and authoritative voice and to follow any noncompliance with a leash correction. Praise was rare, and dogs were seldom, if at all, rewarded for responding correctly. The training mantra seemed to be "command and correct." Later I learned the mantra was actually "command, correct, praise," which didn't make too much sense to me. Praising after a correction seemed more like an apology rather than a training philosophy. Maybe they meant praise if the correction worked.

I came with a notepad, and I started scoring what I called "command:correction ratios," and I calculated the percentage of responses that were corrected. I focused on four handlers. As would be expected, most corrections occurred while heeling and especially after sitting. Dogs received 2.7 leash corrections following each Heel command, and nearly 90 percent of Sits (during heeling and after recalls) were followed by one or more corrections.

That evening, Mimi and I had a long discussion regarding what exactly the dogs were doing "wrong" that required so many corrections. Once you know the problem and which questions to ask, solutions and answers usually follow. Obviously, the dogs had very poor comprehension of basic commands and certainly would benefit from being taught ESL. The prime reasons for the many corrections following Sit appeared to be that the dog didn't sit quickly enough and Sits were crooked or out of place (too far away). The numerous corrections when heeling were a bit of a mystery.

I went back the following week determined to understand the heeling corrections. I brought along sheets of paper depicting a handler's right and left footprints, and marked an X showing their dog's position each time the dog was corrected. When heeling, 45 percent of corrections occurred when the dog was too far ahead, forging or pulling; 35 percent when lagging too far behind; 15 percent when too wide; and 5 percent when too close and/or bumping the owner.

Unbelievably, that second week, dogs received 2.8 corrections per Heel command and exactly 90 percent of Sits were corrected. Despite so many leash corrections, behavior was not improving, as evidenced by no reduction in the number of corrections. The leash corrections were not working, nor were they "correcting" specific problems. In other words, the leash corrections *were not corrections!*

It is extremely difficult for dogs to learn what's "right" if there are no instructions and if the only feedback is a leash correction for getting it "wrong." This epiphany led me to develop *verbal* corrections. I asked the class's trainer, who was a friend of mine, if I could work with the four handlers. Obviously, "forging" and "lagging" are antidotes for each other, and so in a subsequent class, I taught the four handlers how to teach Hustle and Steady (that is, to put forging and lagging on cue). In a single short session, the number of leash corrections following each Heel command dropped from 2.8 to 0.25, a reduction of 91 percent.

Dogs require clear and precise instruction for each exercise, and they require specific guidance for each problem. Too often, a single type of punishment, such as a leash correction, is used as a one-size-fits-all, attempted solution.

"Training" a Human Guinea Pig to Sit

Years ago, during an East Coast lecture tour, Dr. Katherine Houpt asked me to pop in and talk to the veterinary students at Cornell. At lunch, I asked her whether she could find me a student with an exceptionally sound temperament for an onstage demo. I met the young man and explained the demo might be a little strange or scary, but he would come to no harm. He was game. So I taught him a "safety signal" and said he could opt out at any time (without affecting his grade).

That afternoon, I invited him onstage where I had arranged ten chairs in a row. I politely requested him to sit down. As he walked over and was about to sit, I grabbed him by the lapels and screamed in his face, "*Noooooooooo!*" Then I commanded, "Sit!" again. This time, he moved tentatively, and as he started to sit, I twirled him round and landed him on the floor with a judo hip-throw, much the same as alpha rollover. His eyes got big, and he gave me the "safety signal."

I asked him, "Would you please tell the audience what you have learned?"

He replied, "Dr. Dunbar, you're a jerk!"

I think he described my training demonstration better than any dog could. I offered my apologies and asked, "What do you think I wanted you to do?"

He replied, "I haven't a clue. You asked me to sit, but both times I tried to sit, you just got angry."

I said, "Please bear with me and I'll show you."

I walked to the chairs and said, "Please, Sit." Then I motioned with my hand for him to sit on the chair *at the end of the row*. As soon as he sat, albeit still tentatively, I pulled out a couple of twenty-dollar bills from my wallet as a reward. I asked him again, "So what do you think I wanted you to do?"

This time, he replied, "To sit on *this* chair."

Then I explained to the audience, if you don't give clear instructions, yet try to train by punishing a dog (or student) for predictably making mistakes, not only will it take ages to teach them what you want them to learn, you'll most certainly harm your relationship. That is, the dog will learn their person is a jerk and not to be trusted.

Just imagine how long it would take, and how many physical grabs it would require, if there were a hundred chairs on the stage and I didn't specify *which* one? This is not much different than when we don't teach dogs *where* to eliminate or *what* to chew.

Lack of clear instruction is the number-one reason why punishment fails. Remember, the veterinary student willingly agreed to participate in my demonstration, he was willingly compliant, but he still failed. Why? Obviously, because of the trainer and the "training" technique. My punishments certainly conveyed to the student that he was doing something "wrong," but not *what* it was nor *how* to get it right.

The power of specific verbal instructions, like "Chewtoy," "Bed," and "Sit," becomes obvious. Not only do they tell the dog to stop doing whatever they are doing, but they provide the instruction for what to do instead, and for which they will be handsomely rewarded.

As a side note, I bumped into this student some ten years later. I was looking at the referral statistics for SIRIUS, and I noticed that a small veterinary clinic in North Oakland, which had never sent us a puppy in ten years, was suddenly referring a dozen puppies a month. When I popped in to say thank you, I discovered that my student guinea pig was now a veterinarian and spreading the good word about puppy classes.

Misconceptions about Punishment

When learning theory made its transition from shocking rats in laboratory research studies to dog training, most trainers assumed that the terms *aversive* and *punishment* were synonymous. That has led to two huge misconceptions about punishment that continue to this day: (1) That *punishment must be aversive to be effective*, and (2) that *all aversive stimuli are punishing*. Both assumptions are incorrect!

In fact, many behaviorist and psychological definitions of punishment make no mention of discomforting, unpleasant, aversive, or painful stimuli. Instead, most definitions state that punishment is a "stimulus," "stimulus change," "procedure," "event," or "consequence" that causes a decrease in the frequency of a behavior.

As you have learned, punishments and rewards are not defined by their *nature* (whether they are unpleasant or pleasant) but by their *effect on behavior*. Strictly speaking, anything (even if enjoyable) that decreases behavioral frequency is a "punishment," for example, using the command Sit to decrease or stop jumping up. Conversely, anything (even if unpleasant) that increases behavioral frequency is a "reward," for example, when shouting at a dog increases barking, biting, or running away.

Here is one simple, behavioral definition of punishment: *A stimulus that inhibits an immediately preceding behavior, causing it to be less likely to occur in the future in that situation*. This definition makes no mention of pain or discomfort. Of course, the stimulus *could be* aversive, painful, or scary, or it *could be* nonaversive, pleasant, and welcomed. Both giving a leash correction and instructing Steady would each be defined as a "punishment" *if* they both resulted in a dog pulling less while heeling on-leash.

Aversion and Punishment Are Not Synonymous

One reason for the misuse and excessive use of aversive stimuli in dog training is because of our everyday understanding of what punishment means in human society. Within our legal system and dictionaries, and sometimes in education and psychology, *punishment* is defined as the following: pain, suffering, or loss inflicted on a person as a retribution for wrongdoing; severe, rough, or disastrous treatment; and an established penalty for a crime or offense, which might include deterrence, rehabilitation, incarceration, and incapacitation. A *reprimand* is usually defined as expressing disapproval or admonishment, a sharp reproval or severe rebuke, or scolding.

In other words, in society, punishments and reprimands usually relate to our sense of moral or ethical justice. They are ways to right a wrong, and so people feel strongly that punishment *should* be unpleasant or painful. Most definitions of punishment focus on two aspects: (1) the reasons for punishment, that is, the commission of a crime or an instance of intentional hurtful behavior, and/or the omission of an expected or required action (such as not helping or silently allowing a wrong to occur), and (2) the nature or means of punishment, that is, the appropriate penalty that fits the crime, whether that's offering an apology, paying a fine, incarceration, or corporal or capital punishment.

Punishment Need Not Be Aversive

Just to reiterate, in dog training, a punishment can be any type of stimulus that reduces the frequency of a behavior. That stimulus *does not need to be aversive*. In fact, the most effective punishment bar none is *verbal instruction and guidance* — even a single word can prevent, reduce, or eliminate so many undesired behaviors.

Aversive Stimuli Are Not Always Punishing

People also assume that all aversive stimuli make effective punishments, but nothing could be further from the truth. Just because a stimulus or an action is aversive or painful doesn't mean that it will effectively inhibit or eliminate specific behaviors, for example, the leash "corrections" in the obedience class I audited.

By definition, *effective* punishment causes a progressive decrease in the frequency of the targeted behavior until, eventually, the behavior is eliminated. As such, we would expect a concomitant decrease in the need and use of punishment until there is *no need*. Effective punishment eliminates the need for more punishment.

Is that what we see with aversive corrections? Even from casual observation, we see people repeatedly jerking the leash during class after class and on walk after walk. *Continued and excessive use advertises the ineffectiveness of physical correction.* This shows they are not working as intended.

Moreover, if aversive stimuli do not rapidly reduce the frequency of misbehavior, leading to its elimination, then they *cannot be reasonably defined as "punishment."* They are merely... aversive. All they do is inflict discomfort and pain.

Confusing the *Nature* and *Effect* of Punishment

Although the terms *punishment* and *aversive* are clearly *not* synonymous, to this day, many people continue to confuse the *nature* and *effect* of punishment. Any stimulus *intended* to be a punishment may be defined by its nature (whether it is aversive or not) and by its effect (whether it works or not, since punishment is defined by its effect on behavior). The table below will help clarify the differences.

		IS IT PUNISHMENT? (IS IT EFFECTIVE?)	
		YES	NO
IS IT AVERSIVE?	YES	Aversive Punishment	Aversive Nonpunishment
	NO	Nonaversive Punishment	Nonaversive Nonpunishment

The first time I heard someone say, "I hate seeing all these *aversive punishments*," I thought, *But you can't be.* Only very few stimuli or procedures qualify as *aversive punishment* because they have such a fleeting existence. When they do their job (by eliminating the undesirable behavior), they are no longer required, and so are no longer used.

Instead, what we see are lots of *aversive nonpunishments.* These are aversive stimuli that are *intended* as punishment, but since they do not fulfill their definitional duty of reducing undesirable behavior, they are actually "nonpunishments." Their frequent use continues because they are ineffective, and undesirable behavior continues.

Additionally, we see lots of *nonaversive nonpunishments*, for example, begging, beseeching, imploring, pleading, and nagging. Nags are repeated over and over with the intention (or hope) that the dog will comply, but they don't. While nagging and pleading can be an irritant for some dogs (and people), they are not aversive, in the sense of being scary or painful. Well, painful to the ears maybe.

That leaves us with the fourth quadrant: *nonaversive punishment*. This is what we need — *a stimulus or procedure that accomplishes the goals of punishment (reducing misbehavior) but without causing fear or pain*. The most obvious example: *verbal instruction*. Way back in the eighties, I used the term "instructive reprimands," but because the word *reprimand* is also tainted, and people think they have to shout or wag an admonishing finger, I renamed these "single-word instructions."

I consider nonaversive punishment the most-important and most-neglected aspect in training animals, teaching people included. When we simply use verbal instructions and guidance, we change our entire outlook about training. For the last century, we have been duped into believing that aversion is necessary, largely because of laboratory research, wherein computers could only use shock to inhibit behavior. We don't need to. We can use our voice, and when we do, we find it is far more effective than using any physical means.

Six Criteria for Aversive Stimuli to Be Effective Punishment

Why do so many aversive stimuli fail as *effective* punishment? Basically, because for aversive stimuli to reduce and eliminate unwanted behavior, their *application* must adhere to six strict criteria. In laboratory research, it's relatively easy for computers to administer aversive stimuli effectively to punish captive rats, but it's nearly impossible for people to fulfill all six criteria in dog training. Please do not be seduced by the ease of grabbing a dog, jerking a leash, or pressing a button on a shock collar. Using aversion to effectively resolve problems is an extremely exacting and time-consuming process that requires a colossal skillset. The application of aversive stimuli must meet these six stringent criteria to be effective and work as punishment:

1. Punishment must be *immediate.*
2. Punishment must be *consistently* administered.
3. Punishment must *"fit the crime."*
4. Dogs must *understand* rules, requests, and commands.
5. Trainers must *warn* prior to punishment.
6. Punishment must be *instructive.*

These six criteria are quite an ask, so let's go through them one at a time. Usually, the *immediate* and *consistent* administration of punishment are the deal-breakers for so many owners and trainers. And for dogs, the lack of instructiveness of aversive stimuli is the doozy.

1. Punishment Must Be *Immediate*

Numerous studies have been conducted on the immediacy of punishment. To be maximally effective, punishment must be delivered within a split second following the undesirable behavior. A delay of just one second reduces the effectiveness of punishment to approximately 60 percent, after two seconds to 20 percent, and after three seconds, the "punishment" has zero inhibitory effect on behavior.

To put this another way, a delay of just two seconds necessitates that the punishment must be *five times as severe* to have the same effect as an immediate punishment. Bad timing is an absolute disaster for aversive punishment. After three seconds, it simply teaches a dog that their owner is prone to unpredictable fits of "nasty." Obviously then, reprimands and punishments are essentially incomprehensible to a dog when delivered hours later, since a dog simply cannot make the association between an unpleasant event when their person arrives home from work and something the dog did that afternoon.

Many people interpret their dog's behavior when being yelled at as evidence of guilt; they say, "He *knows* it's wrong." But more likely, flattened ears, a tail between the legs, and rolling over are appeasement gestures in response to the person's agitated demeanor and angry tone.

After just a few arriving-home outbursts by their owner, the dog may become increasingly apprehensive and agitated as the late afternoon wears on, and they may act fearfully as soon as they hear the car in the driveway.

From the dog's point of view, the person's homecoming behavior is unpredictable; sometimes they're happy, sometimes they're not. Eventually, the dog may work out predictors of whether their person will get angry: the *evidence* of the "crime," chewed remnants or feces on the floor. But even then, we cannot assume that the dog "knows" their *behavior* is wrong.

Imagine the situation from the dog's perspective. They are sleeping peacefully and wake with a sudden urge to defecate. It doesn't matter the reason: Maybe they have an upset stomach, or maybe their owner didn't take them outside to defecate before leaving the house. Now the dog is home alone and without a key to the toilet. What's a dog to do? They have to go.

I suppose if the owner set up a camera, they could review video of the dog's behavior to see whether the dog was clearly stressed — pacing back and forth, scratching at the front and back doors trying desperately to get to the toilet. Or, did the dog look straight into the camera lens, flip it off with four front digits, and take a dump in the middle of the carpet? That might clarify the dog's state of mind.

Bad timing not only renders punishment useless for inhibiting undesirable behavior, but it also can inhibit desirable behavior. For instance, consider recalls in a park. Sometimes a person will repeatedly scream over and over, "Come HERE!" or they will keep cooing and bribing with biscuits. Either way, they get little response, and when their dog eventually comes, they let out all their frustrations. They grab the dog, snap on the leash, and scold the dog all the way home. What does the dog learn? Obviously, that coming when called results in anger and reprimands, so in the future it would be smart to delay recalls for as long as possible.

Always praise after your dog comes when called, no matter how long it takes. And when your dog eventually comes, don't leave the park, but instead start practicing park recalls, since now your dog is ready to be taught.

Appreciate and enjoy life with your dog because, one day, sadly, you'll no longer have that dog to take to the park. *Those whom we grieve tomorrow are alive and well today.*

2. Punishment Must Be *Consistently* Administered

For punishment to be effective, a dog must be immediately punished after *every* transgression. Every single one! This is simple in theory but next to impossible for people to accomplish in practice.

If someone fails to punish their dog after a single misbehavior, the dog will begin to discriminate instances when noncompliance or natural doggy behavior is never followed by punishment. For example, the dog may learn that they may safely chew and dig and bark and otherwise act like a dog when their person is absent — *separation joy*!

Examples include an owner being *physically absent*, such as in a different room, in the shower, or away from home; being *physically present but functionally absent*, that is, when the dog is off-leash, or when the owner's hands are occupied with a mobile phone and a cup of tea; and being *physically present but mentally absent*, that is, when the owner's brain is preoccupied, checking email, watching the telly, reading a book on dog training, daydreaming, stargazing....

Let's consider a hypothetical Yorkie who's developed a bad habit of peeing in their owner's bedroom. When their person is home, let's say the Yorkie is punished every single time immediately after they pee! The *immediate* and *consistent* punishment teaches the Yorkie not to pee in the bedroom — *when their owner is home*. But the Yorkie quickly figures out that they are never punished when their person is away from home. Even effective punishment can complicate and exacerbate problems by creating *owner-absent* behavior problems and *off-leash* training problems.

3. Punishment Must *"Fit the Crime"*

I have seen people become totally unhinged when their pup pees or poops indoors. Human histrionics and temper tantrums for a *potty-training accident*? Do we act the same way when babies poop in their diapers? Or when older men become incontinent? I've witnessed people grab the pup by the scruff and rub their nose in feces. Dog training books used to recommend this "technique." What on earth are we trying to teach the pup? For the pup to solidly tamp fecal products into the carpet pile each time?

Minor problems, especially puppy misdemeanors, require further guidance, not scolding and punishment. Yes, unexpected feces in the house can be upsetting, particularly depending on its location, but we shouldn't take our feelings out on our dog. Peeing and pooping indoors is a very different level of "crime" compared to barking, growling, and lunging at people or other dogs. Even so, there's no need to get physical there, either.

4. Dogs Must *Understand* Rules, Requests, and Commands

I have stressed this point throughout this book: Before even considering punishing a dog for misbehavior or noncompliance, first check that the dog understands ESL. Prior training and testing are requisite. Without knowing what's expected, dogs will no doubt improvise and break rules that they don't even know exist. *There is no better way to prevent misbehavior than teaching dogs how to behave as desired from the outset.* Moreover, teaching ESL and offering verbal guidance is the best way to eliminate existing problem behaviors. In a sense, *the most effective punishment is lure-reward training and verbal instruction.*

5. Trainers Must *Warn* Prior to Punishment

Not teaching and issuing an understandable warning prior to administering punishment is both inane and unkind. *Inane*, because with no prior warning, the dog will never learn the meaning of the warning and so will likely continue to misbehave and be "punished." *Unkind*, because with no warning, the dog has no opportunity to *avoid* an aversive punishment, and that isn't fair.

The importance of prior warnings came from laboratory research on caged rats. When I first read this study, it had a sickening effect on me, but it later made me realize the importance of clear verbal instruction prior to the task and additional verbal guidance afterward.

Basically, in the study, rats were shocked at unpredictable times and without warning, causing them to respond frantically in sheer panic to *escape*. After several unpredictable and prolonged shocks, many rats became extremely inhibited and "shut down." Some became helpless and no longer even attempted to escape the ongoing pain; instead, they lay down on the electrified grid, even though a safe haven (a shelf) was available.

It is one thing to use aversive punishment to inhibit undesirable behavior, but to administer long-term shock without prior training and warning is just beyond the pale. This experiment reminded me of the puppy-shocking experiment that gave rise to the myth of a so-called "fear imprint period."

I really felt for the rats. I have always loved rodents, especially mice, rats, hamsters, cavies, and capybaras, and even gophers (in other people's gardens). But I am so glad that I read that study. It prompted me to read many other studies, and it inspired me to question whether punishment ever needs to be aversive. It started my quest for effective, nonaversive alternatives.

In subsequent experiments, if a buzzer forewarned the rats that the metal mesh floor was about to be electrified, they quickly learned to *avoid* the shock by immediately hopping onto the ledge the moment they heard the buzzer. But they still didn't know when it was safe to come down, and so they kept testing the floor with their whiskers. However, once a "safe cue" was added, the rats spent a lot of their day hopping onto the ledge when the buzzer sounded, and hopping down again following the safe cue, with no discernible signs of stress.

These experiments, though unpleasant, reveal the extreme importance of *always* warning prior to administering any kind of aversive punishment. Hopefully, the memory of the rats in that one study will continue to inspire change that makes life safer and more enjoyable for all the animals we train.

Once animals understand what it signals, a warning enables animals to avoid the pain. In the studies, after just one warning and one shock, *the frequency of shock went from one to zero in one trial.* Yes, aversive punishment *can be* stunningly effective when automatically dispensed consistently and immediately after a warning — by a computer in a laboratory on caged animals that cannot escape.

But does punishment work that well when *people* try to punish animals that are not restrained or confined? Dogs can run away. And so will people in similar situations. Children play truant or run away from home, spouses divorce, and employees resign.

I warn dogs all the time. In a sense, that is one function of a dog's formal name. When said prior to any command, it forewarns the dog that I *will* follow up, by calmly and persistently insisting until they comply, and then repeat the exercise until they comply following a single command.

6. Punishment Must Be *Instructive*

This is where *aversive* punishments crash and burn. Any hope of instruction (to cease and desist the unwanted behavior) depends on the *immediate* application of punishment.

However, even when aversive stimuli are administered according to the five criteria above as effective punishment to inhibit and eliminate targeted behaviors, *the job is only half done.* Surely, if a dog is punished for misbehaving, the dog still needs to be taught and trained to behave in the way we would like, for example, to heel closer and quicker, to sit quicker and straighter.

Consequently, *even effective punishment is insufficient* because we still have to get the dog back on track. This, of course, is where our voice reigns supreme. Even a single word can convey the two vital pieces of information: Stop what you are doing, and do this instead.

The Importance of Educating All Trainers about Other Options

A lot of dog trainers ask me why I discuss aversive punishment at all, since I never recommend its use. Largely because the terms *punishment*, *negative reinforcement*, and *negative punishment* have crept into dog training books and, more disturbing, pervaded our interactions with people. I think it's important that people understand these techniques, acknowledge their many constraints for use in real life, and learn what's missing in training and education, so we can focus on providing more-effective, nonaversive alternatives.

Additionally, rather than only preaching to the choir, I do my best to encourage *all* trainers to consider every option to find the easiest, quickest, and most effective techniques, and especially so *if trainers use aversive techniques*. I think I can make a huge difference in a dog's quality of life by teaching trainers how to use a punishment more effectively and how to use a food reward more effectively. If trainers choose to use aversive techniques, of course, I want them to know how to increase effectiveness. Remember, the introduction of just a warning buzzer eliminated the need for more than one shock in lab experiments. Similarly, when food lures and rewards are used effectively to teach ESL, any aversive stimulus becomes unnecessary.

Lure-Reward Training versus Effective Punishment

Throughout this book, I've maintained that, of course, we all want to decrease misbehavior and noncompliance, but that *aversive* punishment is *unnecessary*, and even when effective, *insufficient*. Also, I objectively state that lure-reward training is an easier, quicker, and more effective way to accomplish the same goals. Let me try to justify my second statement.

Lure-Reward Training Is *Easier*

Lure-reward training is comparatively simple, especially in terms of feedback. It really is next to effortless and requires minimal brain power to say, "Thank you" or "Go play." On the other hand, to use aversive stimuli as effective punishment requires adherence to six stringent criteria and a considerable physical and mental skillset.

Additionally, lure-reward training becomes easier as training progresses, whereas punishment becomes more difficult. The more we reward, the more desired behaviors increase in frequency, which makes them easier to reinforce, and this creates a wonderful positive feedback loop. Eventually, desired behavior squeezes out misbehavior, which makes reward-based training the best way to reduce and eliminate the need for punishment.

Effective punishment, on the other hand, decreases the frequency of misbehaviors, making it more difficult to complete the program. To eliminate undesirable behavior entirely, dogs must be "caught in the act" and punished in every instance, even when the owner is physically, functionally, or mentally absent.

Lure-Reward Training Is *Quicker*

To reduce and eliminate misbehavior effectively using punishment, the trainer must punish the dog for *each and every* instance until desirable behavior becomes the last behavior standing. But let's just consider: In how many different locations might a Yorkie choose to pee in a bedroom alone? On the floor, on the carpet, in a closet, on a sock-and-underwear shelf, in a dirty clothes basket, in slippers, under the bed, on the bed, on the pillow, and so on. And how many times in each place? That's a lot of times in a lot of places — an *infinite* number of times and places.

Therefore, for punishment training to be effective, the trainer needs to be present to punish the dog for an *infinite* number of improvisations, which of course requires an *infinite* amount of time.

However, when teaching a Yorkie *where* to pee, *what* to chew, *where* to lie down, or *when* to shush — there is *ONLY ONE RIGHT WAY*, which requires only *a finite* amount of time to teach. Since *finite* is shorter than *infinite*, lure-reward training is quicker than punishment training.

Lure-Reward Training Is *More Effective*

For punishment training to be effective, the trainer must adhere to six stringent criteria, which is next to impossible outside of a laboratory setting, wherein computers train captive animals. However, these same criteria are either unnecessary, inadvisable, or embedded within the process of lure-reward training.

Instruction is the essence of lure-reward training by teaching the meaning of words, increasing their learned relevance, and then verbally offering initial instruction and ongoing guidance. Dogs commit no "crimes" as we teach them. Single-word instructions can be used extremely effectively to *teach* dogs to cease and desist *and* instruct them what to do instead. And poof! The undesirable behavior is history. So easy, so simple, and so amazingly effective.

Also, the *immediacy* and *consistency* of feedback is not a critical issue. In fact, throughout this book, I have emphasized that the longer we delay giving a reward, the greater the number and the longer the duration of reinforced responses.

Similarly, inconsistency *increases* the power of reinforcement, anticipation, and motivation, and hence compliance, making inconsistency a virtue. The biggie of lure-reward training is that *consistency is neither required nor advised.*

Given a stellar trainer, punishment training can be effective in a few instances, but lure-reward training is always so much *more* effective, especially for eliminating noncompliance and misbehavior. Lure-reward training is absolutely the way to go.

Negative Punishment and Negative Reinforcement

Behavior is changed by its consequences: by rewarding dogs to reinforce behavior and punishing dogs to inhibit behavior. By now, I hope you have a good grip on how they both work and *how they can both fail to work as intended.* But then along came the notion of terminating either a pleasant situation or an unpleasant situation to change behavior.

The most common example of ending "good times" to inhibit

undesirable behavior is a time-out, whether with dogs, children, or athletes. Meanwhile, the most commonly known example of ending "bad times" to reinforce compliance is human torture.

These concepts can be tricky to understand, and this isn't helped by the ambiguous and perplexing terminology scientists use for these four procedures: positive reinforcement, positive punishment, negative punishment, and negative reinforcement. The terminology confuses everyone, myself included.

What are we meant to understand by "positive punishment"? That punishment is positive? And what's "negative punishment" — extra unpleasant punishment? And isn't "negative reinforcement" just another word for punishment? No, in fact, it's exactly the opposite — this refers to terminating an unpleasant "stimulus," like the pain from torture, to reinforce behavior. Positive reinforcement is the only term of the four that makes remote sense to the lay public.

Ironically, all these terms were coined by a scientist who once wrote a paper with the central goal of establishing scientific terminology that was simple, unambiguous, and easily understandable by a lay audience. Instead, we have mass confusion.

Untangling Terminology

There are two major points of confusion. First, although the terms *reinforcement* and *punishment* are largely self-explanatory, the pair of words was a confusing choice. We must distinguish between "what we do to the dog" and "the effect on behavior": We *reward* a dog to *reinforce* behavior, and we *punish* a dog to *inhibit* behavior. So the pair of words would have been easier to understand if they were either "reward and punishment" or "reinforcement and inhibition."

Second, *positive* and *negative* are ambiguous terms. In common parlance, they mean nice versus nasty, pleasant versus unpleasant, or good versus bad. However, in animal learning theory and dog training, *positive* and *negative* mean to give versus take away, to present versus remove, or to start versus stop, such as starting or stopping a long-term reward or punishment.

Because of my own confusion, I chose the words *posit* (put forward, give, or start,) and *negate* (take away, stop) as a memory aid for what *positive* and *negative* mean in this context.

Our now-familiar concepts of rewards and punishments are represented by the terms:

Positive reinforcement: To give (*posit*) a reward to the dog to *reinforce* desired behavior
Positive punishment: To give (*posit*) a *punishment* to the dog to inhibit undesired behavior

The two new terms for ending a pleasant or unpleasant stimulus are:

Negative punishment: To stop (*negate*) good times to *punish* the dog and inhibit undesired behavior, for example, by using a time-out
Negative reinforcement: To stop (*negate*) long-term pain to *reinforce* desired behavior, for example, by using torture to get a confession

Now, let's evaluate the effectiveness of negative punishment (using a time-out) and negative reinforcement (using a forced retrieve) for modifying dog behavior, while the obvious goal is to devise alternative, *nonaversive* applications that use the power of these techniques without causing pain.

Negative Punishment: Time-Out

The process of negative punishment involves abruptly ending an enjoyable activity or scenario to suppress undesirable behavior. In dog training, these techniques are fairly benign, such as using a time-out to inhibit inappropriate play. With people, the most familiar examples are parents grounding their children for staying out too late, spouses giving the silent treatment when the other is rude, detention in schools, incarceration for breaking the law, and more recently, "ghosting" people online. There are many excellent examples in sports, such as time-out penalties in ice hockey, red cards in football, and fouling out in basketball. In essence, enjoyable activities, freedom, and sometimes communication and affection are abruptly withdrawn to inhibit undesirable behavior.

The power to modify behavior comes from the *abrupt and huge transition* from good times to not-so-good times. However, in addition to evaluating the speed (effectiveness) of the transition, we also need to zoom out to evaluate what happened prior to the transition and what happened after the transition, especially since the preceding and following periods are often lengthy. Looking at the bigger picture uncovers a lot of problems. Also, as always, we must consider the proceedings from the dog's perspective as well as the trainer's.

During a dog training class, when a dog barks too much, has a pushy play style, or exhibits amorous intentions, the trainer may decide to give the dog a time-out. The question is, how well does it work?

The Transition

An *instantaneous transition* from the dog enjoying playtime to being confined, or put on-leash, is essential for the dog to make a connection between their transgression and the resulting consequence — the time-out. However, this seldom happens. For example, when a dog amps up and barks in class, some trainers walk over silently, take the collar, lead the dog to a dog crate, pop them inside, and shut the crate door, where the dog is left alone until deemed sufficiently calm and quiet to rejoin the group.

The procedure violates several of the criteria for punishment to be effective: no warning, a long delay between "crime" and punishment, no instruction, and a lengthy time-out that is overkill as a consequence. Although *negative* punishment is considered to be nonaversive, *it is punishment* for the dog, and so must still adhere to the criteria for effective punishment.

With lengthy transitions, rather than associating confinement with the infraction, that is, *the dog's undesirable behavior*, the dog will more likely associate confinement with events that were closer in time to the crate door closing — that is, the *trainer's* behavior. The dog may learn to mistrust the trainer, and in the future, avoid their silent approach and collar take, and resist entering the crate.

An effective transition should take less than half a second. The only way for this to happen with a time-out is for the trainer to use their voice to instantly inform the dog of their infraction, in the same manner as a referee using their whistle.

Before the Transition

What happened before the time-out? Usually, uninterrupted dog-dog play without sufficient verbal guidance, which allowed the dog's behavior to escalate to a point where it was deemed out of control. In sports, a good referee wouldn't allow this to happen and spoil the game. Instead, as the pressure of the game begins to boil, there would be whistles and warnings for lesser infractions.

Trainers and owners can avoid the need for time-outs by being good "referees," by providing ongoing verbal guidance, especially Steady, Settle, and Shush, and by integrating numerous training interludes into play to periodically calm the dogs.

After the Transition

What happens after the dog is put in time-out? The dog is confined for an undetermined length of time, and the dog receives little to no instruction or guidance for what to do (settle and shush) to earn a reduced sentence.

A trainer might say, "I'm totally positive. I only use negative punishment. I only remove freedom and privileges." But what might the dog say while sitting in a crate watching all the other puppies having a rollicking time? *What does she mean? This is positive punishment. I'm in prison! I shouldn't be in solitary confinement. I'm a social animal.*

"Shutting the crate door" is perceived as "ending good times" by the trainer, but as "the beginning of bad times" by the dog. So often in training, the dog's and trainer's perspective of the proceedings are polar opposites.

Now, since the dog perceives confinement as punishment, what better opportunity for the trainer to immediately release the dog from confinement the instant they settle down quietly? This, of course, would be a *nonaversive* application of negative reinforcement. Even better, to speed up an early release, the assistant could lure-reward train the dog to Settle and Shush while confined.

Owners do not relish seeing their dog singled out for confinement. They came to class to learn how to train their dog and *control their exuberance*, not see their dog penalized for it. Both owner and dog would value one-on-one personal instruction by a trainer's assistant, and what better time to start than when the dog is confined, calming down, and becoming focused, yet with the clamor of other dogs playing all around. Maybe instead of using a crate as a prison, put the dog on-leash instead, so there is a person at the other end of the leash to teach the dog. Or offer one-topic troubleshooting classes specifically to teach unruly and noisy dogs to be calmer and quieter, so that a single unruly dog does not ruin a basic manners class. Such classes fill up as soon as they are scheduled.

My extreme pet hate is any lengthy training dead time in a dog training class, such as when a dog is confined to a crate without instruction, or when owners sit in chairs with their dogs on-leash while a trainer lectures about theory, or what they're *going* to do, rather than guiding owners while they do it.

I use time-outs primarily in two situations: (1) *A game stops when game rules are broken*, and (2) *training stops when dogs cease to be engaged.*

For example, if there is *any* skin contact by the dog's teeth when playing tug, or any body contact in a game of tag, I say, "Penalty!" For example, I would point to the penalty box and little Oso had to immediately walk to the bottom step and lie down until released. Immediate transition! Meanwhile, I would play with Phoenix and accentuate and advertise the good times we were having. "Hey Phoenie, wanna have one of Oso's treats? Yum, Yum!" Once I saw I had Oso's undivided attention, I'd put him through a few rapid position changes, ask him if he wanted to play tug, and then repeat Off / Take it / Thank you a few times, to check that he could take the tug toy without touching my hands.

Similarly, when I have really tried my best to engage a dog, but it is just not working, I say, "Finish," give an ASL sign that indicates that training, which should be the best thing in the dog's world, is *over*. I'll turn my back on the dog and laugh, and then have an especially good time doling out oodles of food rewards while training another dog. Again, an immediate transition.

Time-outs in animal training are often ineffective because of the lengthy transitions between the "crime" and punishment, but the *nature* of their application is usually fairly benign, aside from the training dead time, which I do consider to be a "crime."

During dog-dog play, I never use a crate for punishment. Instead, I usually say, "Rover, Sit!" and then have a calm chat with the dog: "What on *earth* do you think you're doing? You know better than that. Now let's bring it down a notch. Alright, Rover, Go play. There's a gooood dog. Keep chill. Gooood boy." With ample verbal guidance and loads of praise, eventually time-outs become unnecessary.

Similarly, aside from sports, time-outs with people often fail for two reasons: first, because the punishment is delayed excessively, for example, when grounding, giving the silent treatment, ghosting, and especially legal incarceration. And second, again, because of the very lengthy communication/education dead time during the time-out, which is the ideal opportunity to talk and explain hurt feelings and try to resolve the issues. That way, everyone can get back to life as normal as soon as possible. Remember, just as we love our dog, even if their behavior is occasionally irksome, we love family and friends, even if we are occasionally upset by their behavior. Moreover, holding hurt, anger, or a grudge hurts us more. Never forget: It's the behavior we dislike, not our dog, our spouse, our child, or our friend.

Negative Reinforcement: The Forced Retrieve

Nearly every application of negative reinforcement in animal training used to be exceedingly harsh, and especially so when misapplied or when things went south. More disturbing, several applications of this technique are surprisingly common within human interactions, especially with children, spouses, friends, employees, students, and strangers, for example, berating and bullying someone until they get it "right." Of course, the most infamous example of negative reinforcement is torture.

Most applications in the animal-husbandry and training world used a wide array of equipment to inflict long-term pain for noncompliance, just so that pain could be immediately terminated the moment the animal complied. For example, a bullhook to get an elephant to move forward; a twitch to get a horse to stand still; and a shock collar to get a hunting dog to sit at a distance or to get a protection dog to let go of the decoy (that is, a person in padded clothing). For years, using an ear pinch, toe hitch, or long-term shock were the default means to get nonretrieving obedience and hunting dogs to take and hold a pheasant's wing, dumbbell, or other retrieval object — that is, a "forced retrieve."

The reinforcing power comes from the *abrupt and huge transition* from long-term pain to no pain. Our goal, then, is to see if we can preserve the extreme effectiveness of the abrupt and huge transition but without inflicting pain or causing distress. I'll use the example of using long-term shock to teach a forced retrieve to analyze the sequence of events.

The Transition

For this procedure to work at all, the pain must be terminated the *instant* the dog complies by taking hold of the dumbbell, bumper, pheasant's wing, or other retrieval object.

Before the Transition

Now, if pain is abruptly terminated immediately upon compliance, obviously, pain must have been initiated sometime beforehand. So we have some really *BIG* questions to ask. Notably: What was the dog doing when the pain started? And how long did the pain last?

What was the dog doing? It's one thing to start punishing a dog when they are already misbehaving and then instantaneously stop punishing when the dog stops misbehaving. But to start inflicting pain when a dog is doing nothing wrong is, well, just wrong. So very wrong. When taught a forced retrieve, most dogs were sitting next to their owners, maybe looking up at their owners, or otherwise minding their own business.

How long did the pain last? Well, that depends on how long it took for the dog to take the dumbbell, and whether the dog ever took the dumbbell. Regardless, usually a long time because most instructions to the dog were physical rather than verbal. For example, opening the dog's muzzle and placing the dumbbell inside.

Often the procedure goes awry. Sometimes the dog never takes the object. Now the trainer has the tiger by the tail, and they are damned if they do, and damned if they don't. If they decide to give up and terminate the shock, they will indelibly, negatively *reinforce abject object refusal.*

Sometimes the dog will bite. Now that is bad news. Not so much the bite. "You hurt my ear, so I hurt your hand" seems a perfectly understandable, interspecies quid pro quo. Rather, since the trainer often stops pinching the dog's ear or shocking immediately following a growl or bite, the trainer will indelibly, negatively *reinforce growling and biting.*

In every instance, *terminating pain is always preceded by initiating long-term pain.* And in most instances, the dog had little if any prior training, and so had no clue what to do to end the pain. The legacy of the long-deceased rats that were shocked without warning dictates that pain should *never* be administered without prior instruction and warning, and especially not under the guise of teaching or training. In the words of a Chinese proverb: *There's no need to use an axe to remove a fly from the forehead of a friend.*

Instead, first shape, lure-reward, or wait-and-reward train the dog Off / Take it / Thank you *on cue*, testing *response-reliability percentages* every step of the way, and using a wide variety of different objects, including chewtoys, bones, balls, tug toys, stuffed toys, pillows, bunches of keys, and of course, dumbbells, hunting dummies, pheasant wings, and protection sleeves.

After the Transition

The cessation of pain provides immediate relief for the dog, but of course, the dog has had to endure long-term pain to feel the relief. The trainer's feedback immediately transitions from long-term pain to no pain — in fact, often to a long-term feedback void, with little if any praise, and barely a thank you. This is what clued me in to the possibility of *nonaversive* alternatives.

Creating Nonaversive Approaches

When I first started to diagnose the *many* problems associated with administering long-term pain, I felt like a mechanic looking at a tractor that had rolled over and the engine was about to explode from "diesel engine runaway": billowing black smoke from burning the oil that was entering the combustion chamber, engine racing uncontrollably, way past maximum RPM, but no way to stop it. Something must give. Just waiting for the "big clunk" when the oil runs out, and the engine seizes.

Animal training should have *never* followed laboratory science down the shocking rabbit hole that promoted pain and especially long-term pain to change behavior. There is never any reason to use pain to *teach* basic manners or compliance. Moreover, with human nature being what it is, of course, pain was begging to be misused. (If you want to learn more, read about the Milgram experiment.)

Lucky for us, we have language, and we can use it in ways that are far more effective than any aversive procedure. Moreover, it doesn't matter how long we wait for a dog to take a dumbbell when using painless procedures; that's what we call "patience." Nonetheless, even when pain-free, to be most instructive, we still shouldn't start the procedure until a dog starts misbehaving, and we must preserve the abrupt huge transition.

As I mention, my first clue came from the abrupt *transition* from "pain to no pain," that is, from pain to essentially a *feedback void.* This seemed extremely odd to me. Why didn't the transition go from "pain to praise"? Especially given the current mantra at the time — "command-correct-praise"? And especially since we're trying to *reinforce* behavior. Shouldn't we at least be thankful when the dog complies? Wouldn't a transition from "pain to praise" be much more powerful than one from "pain to no pain"? I just didn't get it.

If someone had asked me how to teach and reinforce a dog to take an object on cue, I would start by saying, "Rover, Take it," and then *offer a meaty bone.* There ya go. Compliance and reinforcement hand-in-hand and in an instant. I simply would have never dreamt of *stopping pain* to reinforce behavior. For me, "give" just naturally springs to mind, as does "pleasure," as does "give the dog a bone," especially when reinforcement is the intention.

So I wrote down my alternative transition:

Pain → No Feedback → Praise

Then I got it. So obvious. With this longer reinforcement continuum, we have two abrupt transitions in succession and both heading in the same direction, the first *decreasing pain* to zero, and the second *increasing praise* from zero. Essentially, we can communicate the same reinforcement message as transitioning from "pain to no pain" by transitioning from "no praise to praise."

Same Message with No Pain

There's no reason to start with pain, there's no reason to go back to pain, *and there's no reason to even visit the pain half of the continuum.* Instead of transitioning from "pain to no feedback," we can easily communicate the same message in a dog-friendly fashion — *that we're truly happy that the dog took the dumbbell* — by abruptly and hugely transitioning from "no feedback to extreme praise" the moment the dog takes the dumbbell.

Now that sounds familiar. In fact, we call this *wait-and-reward training*, which is amazingly effective because of its immediate and massive transition from "no feedback to praise and reward." When teaching leash walking, the transition is even greater; when the dog starts sitting automatically whenever we stop, our feedback changes from none to "good dog," plus a food reward, plus *the walk resumes*. Essentially, wait-and-reward training is a much more-effective, and decidedly friendlier, *nonaversive* equivalent of negative reinforcement!

Opposites and Extremes

Another negative impact on dog training arose because research scientists thought mainly about *opposites* and *extremes*. First, they considered only *two* feedback options: *either* reward *or* punishment, with no in-between, no words, and no nuance. But obviously, between extreme punishment and supreme pleasure there are thousands and thousands of options.

Second, when coming up with the counterintuitive and scary idea of terminating pain to reinforce behavior, they were extreme and went from a hundred to zero, yet obviously, between "extreme pain and no feedback" there must be thousands of possible graduations. For example, to transition from "moderate pain to no pain," "from very mild pain to no pain," or "from mild pressure to no pressure." Going the whole hog from all to none is unnecessary.

As an aside, the reason why scientists worked with just two types of feedback is because laboratory research study was severely constrained by the apparatus. Computers only had two types of mechanical feedback: deliver a food pellet or deliver a shock. And they had only two options for what to do with pellets or shock, deliver them or abruptly suspend delivery.

However, when scientists presented these as four distinct options, with equal billing, confusion reigned, and it was "training" that went off track, not the dog. Just as words quickly resolve most dog behavior and training problems, hopefully the words in this book can help dog training get back on track.

To further simplify the four proposed options, let's display negative reinforcement (the forced retrieve) and positive reinforcement (wait-and-reward training) on a single *reinforcement continuum*, and negative punishment (the time-out) and positive punishment on a single *punishment continuum*.

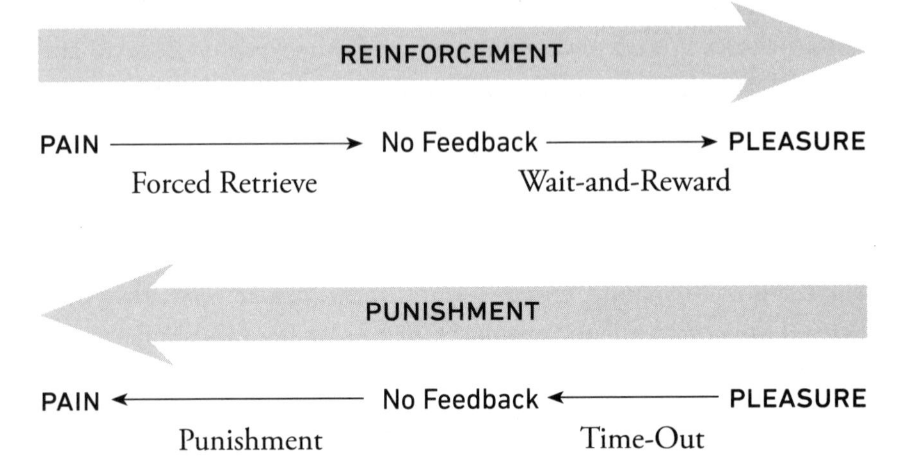

Enormous transitions in feedback are seldom required. For the most part, smaller changes in feedback are more than sufficient to communicate the message: *Praise more when behavior improves and praise less and offer more instructive guidance when behavior worsens.*

Animal-Friendly Negative Reinforcement

Understandably, many trainers and some dog training associations have very dim views about the painful nature of aversive negative reinforcement. However, one cannot ban a principle of learning, especially when there are many useful, nonaversive applications. It's the *effect* on changing behavior that matters: Does it work? How well? How quickly? When it doesn't work well, change to quicker and more effective techniques. It should go without saying that if the *nature* of application is unpleasant or upsetting, change it. Once we dispense with pain, it's win-win.

I have already mentioned one *pain-free* application of negative reinforcement: Wait-and-reward train a timed-out dog by opening the crate door the moment they settle and shush, that is, simultaneously *terminate incarceration* and offer *freedom* in a fraction of a second. Or to speed things up, lure-reward the dog to settle and shush and then release them from the crate.

Occasionally, massive feedback transitions are required, for example when for proofing compliance to formal commands. Repetitive reinstruction until compliance offers another wonderful example of nonaversive negative reinforcement. Instructive, calm, and persistent insistence *starts* the instant the dog becomes noncompliant — that is, when they don't sit following a single formal command — and it *continues* and then abruptly *stops* the precise moment that the dog sits after a single formal command, whereupon we praise profusely and reward by saying, "Free dog" — which is a massive and immediate transition in feedback to reinforce compliance.

Persistent insistence is the verbal alternative to long-term pain. Since verbal insistence is *instructive*, with each repetition, the dog progressively *learns the relevance* of the formal command to Sit.

Teaching Horses to Join Up

Perhaps the most well-known example of nonaversive negative reinforcement is "join up" in horses. Many horses develop a fear of people, but during a single short session, they shed their fear and "join up." In a nutshell, when a person approaches a scared horse, or merely looks at their hindquarters, the horse will move away. In a round pen, the horse cannot escape, and so they circle. And circle and circle and circle... until eventually, the horse signals that they give up the gambit by lowering their head, whereupon the person *turns away and walks away*. Most horses then approach and walk beside the person, or join up. It's fascinating to watch. Of course, the join-up process could be considerably accelerated with a few alfalfa cubes, each with a couple of drips of molasses.

Some people have criticized the technique, saying that it is inhumane because the horse is pressured, and it is the relief of pressure that reinforces the horse's approach. Yes, absolutely. It's classic negative reinforcement but without causing *physical* pain. And yes, the horse is most certainly pressured. However, the pressure and the horse's mental anguish are neither caused by nor the fault of the horse trainer. The horse has been stressed to the gills most of their life, no doubt caused by lack of socialization as a foal.

The trainer is trying to resolve the problem, and when they usually do, the horse may now enjoy a fear-free and stress-free life around people, for life. The moment they join up, the horse's sense of relief must be immense.

The "stress police" tend to look down on any technique that they think causes stress or fear during training without looking at the bigger picture. Of course, minimizing stress in training is important. However, *not training* would prolong even greater, acute, and overwhelming stress indefinitely. To criticize yet do nothing and so leave the horse to endure a life of fear is not a solution. Certainly, *not training* stressed animals is considerably more stressful than attempting *to alleviate and resolve stress as quickly as possible.*

But as with puppies, the best solution is to encourage handling and socializing of day-old foals to prevent them from becoming fearful of people later in life. Horses are precocious animals, and their impressionable socialization period lasts for just forty-eight hours after birth. This closely resembles Konrad Lorenz's observations of imprinting in newly hatched goslings. Dr. Robert Miller, a veterinarian, cartoonist, and my skiing buddy for over forty years, showed that early "imprint training" in day-old foals produces adult horses that are tractable and friendly toward people, whereas without early handling, horses will *naturally* grow up to be intractable and scared of people.

A Running Commentary

Another big takeaway from analyzing negative punishment and negative reinforcement: Behavior is about transitions. When we zoom out from each transition, we see other transitions that preceded or followed. For example, if a dog is in a crate, at some point they will be released. If pain is terminated, it must first have been initiated. If we zoom out even further, we see a series of transitions.

However, most academic definitions and descriptions of reinforcement and punishment theoretically discuss *single behaviors* and recommended feedback, as if each behavior was static and occurred in vacuo. They use phrases like "capture the behavior" or "reinforce the behavior." This is unrealistic and remote from real life. Personally, I have never seen a "single behavior." Behaviors are pack animals. They live together, roam together, and ... *follow each other in droves.*

Behavior isn't frozen in time for clinical and experimental prodding. Rather, behavior is ongoing and ever-changing, with multiple rapid transitions from one behavior to another, and another, and another ... and at any moment in time, as defined by us, each fleeting transition is unique and qualitatively different.

But it is not even that behavior is an endless series of multiple transitions. Instead, behavior itself is best viewed as one endless transition, always in a state of *rapid and perpetual flux,* which in our eyes is continually and progressively improving or worsening. As such, a few food rewards plus an occasional time-out simply won't suffice as feedback. Instead, our feedback should reflect as many of these ongoing changes as closely as possible.

Obviously, the only way we can do this is by *using our voice* to deliver *ongoing, binary, analogue, and instructive feedback* — by fluctuating our praise and using a wide variety of different instructions as guidance to influence behavior change. Basically, our verbal feedback needs to resemble an old-time radio, running commentary for a horse race or boxing match — responding to every play-by-play action as it happens by offering ever-changing, consequential feedback that reflects each change in behavior.

For example, during dog-dog play, never take appropriate behavior for granted. Praise your dog for "being there" and not barking or bullying. Praise your dog for mannerly behavior, friendly behavior, gentle behavior, and especially, offer verbal guidance for your dog to shush, calm down, or tone down their play when necessary.

Even during a Sit-Stay, your dog's behavior is ever-changing.

Watch your dog, and provide continual feedback based on what you see, mostly oscillating praise with numerous words of instruction and guidance: "Good dog, Rover... that's it, Rove... yooooou've got it — *Sit! Rover, Sit* — gooood boy... gooood dog... back with me, aye?... verrry good Sit-Stay..." You don't want to miss that microsecond eyeball twitch away from you that signals your dog is about to break the stay: "*Watch!*... thank you, Rover... fffooocuss, waatchh me." If they sniff away, or look away, they'll go away.

Obviously, we can't do this all the time. But it's a good exercise, and it's what dogs need. Try it next time when monitoring your dog when off-leash or when walking your dog on-leash. Our dogs need praise when their behavior is exemplary (a lot of the time) or very good (most of the time), and they absolutely require guidance (a lot of the time).

Chapter 15

How Breeding Practices Are Harming Dogs

Genes are vitally important for physical and behavioral health, but present-day selective breeding protocols for dogs require an urgent makeover. This chapter offers my prescription for how to improve them, along with advice for buying a puppy from a breeder.

The biggest issue is that companion dogs die much too young because they are seldom selectively bred for longevity. Dogs were originally bred for working ability and only more recently for conformation and competition. But the careers of working, competition, and show dogs are relatively short: Working and competition dogs wind down their careers as they get older, and the majority of show dogs retire after just a few years.

These days, the most common role for a domestic dog is as a family companion. As such, they should be selectively bred to be companions. When we ask potential puppy buyers what they want from a companion animal, common answers are "good with children," "easy to train," "calm and attentive," and so on. These are all behavioral characteristics over which a puppy's owner has enormous control. However, aside from these obvious traits for companionship, the number-one item on most people's wish list, certainly on mine, is a *long and healthy life*.

People also have control over these two qualities, in particular, dog breeders. And who influences breeders? Their customers. Consumers influence the market, not vice versa, and prospective puppy buyers have a choice: to purchase a puppy that is likely to live to a ripe old age with little if any serious disease, or to purchase a puppy that is likely to die much younger after a number of predictable conditions and diseases. If puppy buyers *insist* that their puppy be descended from a long line of long-lived forebears, breeding practices will change.

Longevity of ancestors is the best predictor of longevity of offspring and their general physical and behavioral health. Since kennel clubs keep records of every puppy that is born, what is desperately needed are records of their age at death to enter into each dog's pedigree. That way, when breeders select a mating pair, they can calculate a "Litter Longevity Index" that may be shared with prospective puppy buyers before making their choice.

Natural Selection Fosters Genetic Diversity

The genetic heredity of dogs has been guided by thousands and thousands of years of *natural selection* and hundreds of years of *selective breeding*. Natural selection embodies "survival of the fittest." Individuals that are best suited to the immediate environment live, while others, who are not, perish. Animals that live and breed pass along the traits that helped them survive to their offspring, thus increasing their chances of survival. In other words, survival of the *genetically* fittest.

Natural selection is nature's fail-safe method to maintain eugenics (good genes and fine offspring). All species are different, and different species fair differently in different situations. The *most fit* do well and stand the best chance of passing along their genes to future generations. The *less fit* do less well, and the unfit usually die. In the wild, the process of natural selection *eliminates nature's genetic mistakes.*

Nature exerts a powerful selection for survival skills. Animals that live longer and breed more have a much greater genetic representation in the gene pool. Individuals may only pass their genes to future generations if they live long enough to breed. Individuals that die young have fewer chances. Individuals that can't or don't mate become the end of their genetic line. In the process of natural selection, death of the genetically unfit is equally as important as survival of the fittest.

Sexual Reproduction Mixes It Up

It always makes me chuckle when breeders say, "We had great success with our last litter, and so we're going to repeat the breeding." I cannot resist asking, "Do you have any siblings?"

"Yes, I have a younger sister."

"And she's just like you?"

"Nooooooooo!"

"Oh. So your parents had you and decided to repeat the breeding and they ended up with your sister!"

Aside from identical twins, sexual reproduction provides a limitless choice of different individuals and is the mechanism that generates genetic variation, since offspring receive half of their genes from each parent. The DNA of each sperm, each egg, and hence each offspring is genetically unique. Then natural selection winnows those individuals to those best suited to the environment, and this influences the gene pool of future generations.

Litters of puppies all display most species- and breed-specific characteristics, but each individual puppy remains genetically unique, so each one also displays a vast array of differences: unique colors, shapes, and sizes, and an enormous variation in behavior, temperament, and working ability. This means there is tremendous variation *within each species, breed, and litter*, which is essential for the survival and betterment of dogkind.

Problems with Selective Breeding

When people selectively breed dogs (artificial selection), the prime directive is *to improve the gene pool along prescribed lines* to increase conformity to a breed standard and to progressively purify the gene pool by *not breeding dogs with inherited diseases and undesirable traits.*

Generally, the effects of selective breeding are much quicker than natural selection. In just a few hundred years, selective breeding has created the many diverse shapes and sizes of dog breeds that we know today, along with an equivalent variation in behavior and temperament.

Extreme unilateral selection for specific characteristics — whether conformation, coat color, temperament, or working ability — often occurs at the expense of other characteristics. As well as improving desirable characteristics, intensive selective breeding often causes unintentional inheritance of undesirable characteristics; for example, increased predisposition to breed-specific diseases and decreased overall health, and hence shorter life expectancy.

To further complicate matters, many undesirable genetic characteristics are not entirely obvious. Often, a dog's "good looks" or "working ability" can be misleading and camouflage the presence of underlying genetic flaws.

This is a huge problem, especially when mate selection is based largely on conformation and/or working ability — as judged in dog shows and trials. When selectively breeding animals, we must be *extremely* particular about which animals we select to breed. Specifically, to improve the gene pool, we must select prospective mates with utmost care, and by *genotype* and not just *phenotype*. That is, by evaluating the genetic constitution of the individual and not just observable characteristics.

Undesired Genetic Traits Are Frequently Passed to Future Generations

Genotype and phenotype often go hand in hand. For dominant traits, only one copy of a gene is required from *either* parent for offspring to express the trait. Consequently, dominant genes can be effectively eliminated from the gene pool by not breeding any dog exhibiting an undesirable trait. However, some inherited characteristics may not become apparent until the dog grows older, much older. For example, genes that cause a predisposition to cancers and other serious diseases.

Meanwhile, undesirable recessive traits require two copies of the gene, one from *each* parent, for the undesirable trait to become apparent. If a dog has only one recessive gene, both the trait and the presence of the recessive gene will remain hidden and so can be passed along to future generations. Dogs with just one recessive gene are usually termed *carriers.* Eradicating recessive genes from the gene pool usually requires eliminating a lot of dogs from a breeding program.

For example, let's say two dogs were bred and just one puppy in a litter of eight displays a recessive trait (double recessive): What do we know about the genetical constitution of the sire, the dam, and the entire litter? Well, using Gregor Mendel's principles of inheritance, we know that: *on average*, 25 percent of the litter are likely to be double recessive and display the trait (although, in this example, it's just 12.5 percent), and 50 percent of the litter are likely to be carriers (single recessive). That is, *on average*, 75 percent of the litter will carry at least one gene, *as do both parents.* Or to phrase it another way, 75 percent of the litter would *appear to be* healthy, yet two-thirds of the apparently "healthy" puppies will be *symptomless carriers.*

Since we don't know which of the apparently healthy puppies are truly healthy and which are carriers, a commonsense precaution dictates that none of the puppies should be bred as adults. Culling is unnecessary, though; these puppies may enjoy an immediate, luxurious career change as celibate companion dogs. However, we *do* know that *both* parents are carriers, and they should never be mated again.

To preserve the health of future generations, it is vital that mates should always be primarily evaluated and selected by genotype, and not only by phenotype, such as looks and behavior.

Genetic Diversity Has Been Reduced

Let's say, at two years old, a good-looking male dog wins best in show in the national arena. Typically, his stud dance card will become fully booked for the next eighteen months with dozens of requests from breeders who want the genes that just won a trophy. Within that time, the breed gene pool will become flooded with the genes from a single dog, and in terms of genetic diversity, the pool becomes a *gene puddle.*

Then what if, at six years old, the stud dog dies of cancer. Unknown to everyone, he had a genetic predisposition to the disease that didn't emerge until early adulthood. But his genetic predisposition is now possessed by many of that stud's offspring, increasing the likelihood that they, too, will develop cancer and die prematurely.

This is exactly what is happening in many breeds. The lack of diversity in many breeds is nothing less than shocking. There are far too many genes coming from far too few dogs with next to no information regarding their genetical constitution in terms of breed-specific diseases, overall health, and life expectancy.

Genetic change within a breed is powered mainly, and surprisingly quickly, via male dogs. Whereas a female's genes are passed along to only a score or two of pups in her entire breeding career, a stud's genetic moiety may be passed along to forty puppies in just one afternoon!

Flooding the gene pool with genes from only one or two males is both unnatural and disastrous for purebred dogs, especially when the stabilizing pressures of natural selection have been removed by human intervention.

Longevity Has Decreased

On average, mixed-breed dogs live longer than purebred dogs, but the gap is narrowing as more purebred problems are showing up in mixed breeds — which is not surprising because so many mixed breeds have purebred dogs in their recent genealogy.

Life expectancy for purebred dogs has been progressively decreasing as the incidence of inherited diseases is increasing, such as allergies, cancer, cardiomyopathy, cataracts, dysplasia, gastric dilation, disk disease, epilepsy, and hypothyroidism. Large and giant breeds, especially, die very young.

However, the trend can easily be reversed by selectively breeding for longevity. In every breed, there are a surprising number of geriatric geezers out there. Most are thoroughly enjoying their sunset years as companions, and they would be only too willing to make a liquid-gold donation to the gene pool.

Three decades ago, at the World Congress of Kennel Clubs in

Bermuda, I first promoted the notion of not mating male dogs until they are seven years old, proven, and still alive. I'm still promoting.

I always like to give cow analogies: Since the time I was in veterinary college, fifty years ago, dairy cows produce 250 percent as much milk. If the farming industry can do this, I am sure that kennel clubs can breed dogs that live longer. There is no reason why life expectancy for all breeds shouldn't be fifteen to twenty years. Bramble, a border collie, lived to be twenty-five; Butch, a beagle, lived to be twenty-eight; Bluey, an Australian cattle dog, lived to twenty-nine. And in May 2023, Bobi, a purebred Portuguese Rafeiro do Alentejo, was crowned the world's oldest living dog after celebrating his thirty-first birthday. The secret to longevity must be obvious to all, name your dog with a B. Be well, Bobi!

Here's another revealing statistic: With improvements in medicine, human life expectancy has *doubled* over the past one hundred years, whereas over the same time, the life expectancy of domestic dogs has decreased.

Solutions for Breeders

When a breeder asks me whether to breed specific individual dogs, my sole consideration is *evaluating the genetical constitution* of both prospective partners and their forebears going back several generations. The most important consideration is the longevity of parents, grandparents, and great-grandparents, which offers the single-best predictor of prospective overall physical and behavioral health and life expectancy of the puppies.

My second piece of advice: If you think there is anything remotely weird or worrisome about the prospective sire and dam, don't breed them. It is not as though dogs are in short supply. Be selective. Very selective.

Don't Breed Dogs That Cannot Breed Naturally

Among domestic species, animals that couldn't breed in captivity did not produce domestic stock, whereas good breeders pass along genes for good breeding to foster future generations of good breeders. Farmers have created livestock that breed like clockwork — *to the day*. However, purebred dogs have seen a huge decrease in breeding ability over the past hundred years. It would be prudent to discontinue trying to breed dogs that have difficulty mating or whelping and not to breed their offspring. All poor breeders can entertain a wonderful career change as happy neutered companion dogs. No shame and no worries about being neutered; they couldn't breed without help, anyway.

What's wrong with helping dogs to mate? Well, whereas there's a myriad of behaviors that are essential for the survival of individual animals, mating behavior is the single-most-important prerequisite for the *survival of the species*. If there were one behavior that nature perfected, it would be mating, and if *that* doesn't work, then other things could be seriously amiss within the dog's genetic constitution.

Being unable to whelp used to be a death sentence for the dam and her puppies. Now *emergency* Caesarean sections save their lives. These days many pregnant females are given elective Caesareans to give birth, but when performed too early, abruptly terminating pregnancy eliminates the hormonal priming of a female's affinity for neonatal amniotic fluids, which occurs one day prior to a natural whelping, and hence, maternal behavior is not bump-started and neonates fair less well, physically and behaviorally.

Courtship Is Nature's Temperament Test

Courtship and natural mating comprise a species-specific, stereotyped choreography of behavioral interaction. Thus, the female is afforded the opportunity to evaluate her prospective mate's mastery of *social* intercourse prior to *sexual* intercourse. Moreover, successful courtship is proof that an excited male dog can get a grip on his impulses and interact appropriately and get the job done without mania, fear, or aggression.

In a sense, courtship is nature's temperament test. Maybe we should heed Mother Nature. Surely, not being able to mate should be the most obvious reason not to mate any dog.

Moreover, despite several thousand years of selection for promiscuity during the domestication of animals, many female dogs remain highly selective in their choice of mating partner and will refuse to mate with some dogs from estrus to estrus, but they will readily mate with other dogs from estrus to estrus. Maybe females know something that we don't.

When in true estrus, the most common reasons for a female to refuse a stud *chosen for her* appear to be that the male is antisocial, asocial, fearful, aggressive, naive, clumsy, or over-the-top in his advances. In other words, courtship and natural breeding weed out studs who don't have sufficient social savvy and finesse, either from experiential deficiency or *inherited predisposition.*

It makes such sense to only mate females that have courted, mated, and whelped naturally and have successfully raised good-sized, healthy litters. As an added benefit, a natural mating helps breeders get the timing right. Many females are mated much too late during true estrus when their eggs have become moribund and so produce tiny litters. On the other hand, a male and female dog acting in combo will time mating to the very day that proestrus transitions to true estrus, which coincidentally is the day of ovulation. Aren't dogs smart? Maybe they should apply to veterinary school.

Delay Breeding Male Dogs until They Are Ten

The questionable genotypes of a small number of show-winning male dogs have become overrepresented in the gene pool of many breeds. So how do we decide which are the best dogs to breed? Looks are obvious, and working ability is obvious, but how do we determine and evaluate a dog's genotype? I would say, by first selecting the ones that are likely to live the longest, that is, those with healthy and long-lived forebears.

Efforts to eliminate inheritable diseases have been championed by breed clubs and a few national kennel clubs. Certainly, when genetic screening becomes affordable and speedy, the puppy-buying public will insist that national kennel clubs make screening mandatory for all purebred dogs prior to breeding. Ultimately, public demand will control supply because people so desperately want a puppy that is likely to live long into their sunset years.

However, rather than combating disease and early death one disease or condition at a time, there is a simpler and cheaper solution. Selecting for longevity and conformation need not be an either/or decision. Select for specific coveted conformation *and* overall health and longevity by simply delaying breeding of male dogs until they are ten years old. By that time, much of their genotype in terms of inheritable diseases will have become apparent.

When writing in kennel club gazettes in the nineties, I stated seven years — the *Seven Star Stud* — but seven is much too early for a male dog to mate. After watching the 2008 BBC documentary *Pedigree Dogs Exposed*, I changed the minimum age to ten.

When I mention this in seminars, many breeders cry: "*Ten* years!?" They protest that the male's sperm count will be too low, or that many dogs don't live that long. Which of course is precisely the point. So many dogs are dead before they are ten.

In terms of the number and viability of sperm, if purebred dogs had a normal lifespan of at least fifteen to twenty years, there would be no reason to question their fertility at ten years. But there are many alternatives. Breeders could collect and freeze sperm from two- and three-year-olds in their prime, but only use it for artificial insemination if the dogs are still healthy and alive to celebrate their tenth birthday. Or if the dog is still alive, the dog could "do it" himself to celebrate his birthday. And then do it again and again because these sperm contain valuable genes to improve the breed.

The ten-year waiting period for mating (or artificial insemination) should apply to male dogs because of their enormous impact on the gene pool, and because it would be unwise to delay mating females. However, prospective dams should be selected that have long-lived forebears.

Stud dog owners are seldom enthusiastic about delaying breeding their champion dog until he is ten years old, but owners of females should insist. Stud selection should be an objectively ruthless process to maximize the quality of a female's puppies in her limited breeding career. Stud selection need be no different from natural selection.

Create a Litter Longevity Index

An immediate way to facilitate evaluation of the genotypes of a potential mating pair is by analyzing the longevity of their ancestors. The beauty of purebred dogs is that their recorded pedigree goes back hundreds of years. Each individual entry includes date of birth and age at mating; by adding another entry — current age, or date of death — this would produce the single-most-valuable, searchable, doggy database for predicting overall health and longevity of all purebred dogs. Also, the database could include the *cause* of death and any lifetime diseases and conditions.

I am sure that crowd-sourced funding would be popular enough to retroactively amend pedigree records over the past decade or two. I've already written my check. Of course, breeder participation would be voluntary, but each dog's age would remain at zero until amended each year.

This way, national kennel clubs can create searchable databases. After entering a dog's name, one push of a button would reveal a single index — the *average* of the current age, or age at death of the dog's two parents, four grandparents, eight great-grandparents, and sixteen great-great-grandparents, and so on. Yes, a lot of the parents and grandparents would still be young. Exactly! They were bred too young, before their genotype was adequately evaluated.

Dog breeders would only have to enter the name of a prospective sire and dam, and the database would analyze longevity going back many generations and then automatically calculate a single index — a Litter Longevity Index — that predicts life expectancy for prospective pups in the litter.

Moreover, once DNA testing prior to mating becomes universal, the inclusion of specific health data would enable the database algorithm to calculate the likelihood that any individual dog carries any undesirable dominant or recessive genes.

Recently, I checked a bunch of dog pedigrees, and for many, every single male dog on the pedigree was mated when they were less than three years old! There's the problem right there. Who knows what genes they are carrying, or how long they are likely to live? Or how young they are likely to die? This is like trying to mate a pig in a poke. It's time we let the cat out of the bag.

A Litter Longevity Index would be a marvelous service for prospective puppy buyers, who would only have to enter the names of the sire and dam to view the life expectancy of the puppies in the litter. Most of the information is in pedigrees already; let's just make it usable for breeders and prospective puppy buyers.

Support for Dog Longevity Programs

I don't think what I am suggesting is rash or out of the ordinary. Rather, it's common sense, and I know there are many like-minded people out there. There already exists a surprising number of individual breed clubs and individual breeders trailblazing the way. However, a Litter Longevity Index program needs to be coordinated by national and international kennel organizations — the keepers of pedigree databases. As always, though, it's passionate individuals and small dedicated groups that make for change. As a tribute, I'll thank two representatives.

For all breed clubs that are trying to make a difference, I would like to laud the Doberman Pinscher Club of America and their "Longevity Program," which awards a Longevity Certificate to any Doberman that reaches the age of ten. The database entry is then updated with the age at death to reflect true longevity. Additionally, the club awards different Bred for Longevity Certificates to any Doberman if both parents live to be ten, or if both parents and all four grandparents live to be ten.

For all dog breeders who want to eventually see white whiskers on their puppies, thank you to Gayle Watkins of Gaylan's Golden Retrievers. As Gayle has said: "The foundation of every decision we make is longevity; we simply do not accept that golden retrievers should live only ten years. We use four tools as we seek longevity: bitch lines with longevity, older sires, high genetic diversity, and outcrosses."

Way to go!

Advice for Buying Puppies from Breeders

People who are looking to buy a puppy from a breeder can ignite and accelerate a massive revolution in the world of dog breeding and husbandry. Many breeders do a great job raising their puppies, but not all. Remember, as a potential customer, you are in the driver's seat. Here are just five things I suggest all potential puppy buyers request of breeders:

1. **Genealogy:** Ask the breeder for a pedigree for sire and dam and any proof that the litter is descended from long-lived forebears.
2. **Early socialization:** Ask the breeder to socialize the puppies to numerous people safely indoors, and safely in the world at large, whether carted or by car. Nothing impacts training more than a puppy's perception of people, which is forged during the first few weeks of life.
3. **Enriched environment:** Ask breeders to expose puppies to every conceivable sort of tactile, proprioceptive, olfactory, visual, auditory, gustatory, and substrate stimuli that they are ever likely to encounter in adulthood, especially sound tapes. Ask breeders to convert the puppies' day room into a jungle gym, with the floor covered with all sorts of substrates, objects, and paraphernalia, so that the puppies' paws barely touch the ground.
4. **Housetraining and basic manners:** Ask breeders to housetrain and chewtoy train the puppies and to teach a few basic commands: Come, Follow, Sit, Down, Stand, Stay, and Rollover.
5. **Proof of development:** Finally, request a weekly video of your prospective puppy's developmental progress. Be polite, but

it's a deal-breaker. If you can't visit your prospective puppy in person, or be sent video proof of your puppy's progress, look elsewhere for a puppy to share their long life with you.

A Special Dedication

I've had the pleasure of sharing my life with many dogs, and I would like to dedicate this chapter to some of them. First to Dune, our American bulldog. Dune was extremely well-trained, and honestly, I cannot recall a single behavior or temperament problem. His temperament was rock solid with people, dogs, children, puppies, little animals, and big animals. Moreover, his entire medical history comprised just two occasions when he scratched his cornea. Because of all that, years ago, I promised him that if he lived to be ten and was still healthy, I would arrange a "boy trip" around the country so that he could spread his precious genes. Dune lived to be twelve, but my work got in the way and we never made the trip.

I also dedicate this chapter to Omaha, my first dog, an Alaskan Malamute who died at five; Hugo, a French bulldog who died at seven; and ZouZou, a Beauceron who stole my heart but died much too young, just after her eighth birthday. The year she died, I promised ZouZou that I would try and do something about the all-too-short lifespans of dogs.

I failed Dune, which was an incredible waste of incredible genes. But I'm keeping my promise to ZouZou.

Conclusion

The Future of Training

Thank you so much for reading *Barking Up the Right Tree*. I hope that your dog is happy with what you've learned, and that you both celebrate with a barkathon or a Jazz up / Settle down session. I hope that daily walks are side-by-side with a bunch of sniff breaks, and that you integrate formal Sits into games of fetch, tug, dog-dog play, all walks, and everything that you do together. That way, at the end of the day, you can both relax during your off-duty times. Please give your dog an affectionate ear scratch and chest rub from me.

Many people view a conclusion as the end; I think of a conclusion *as the beginning*. It summarizes what we have learned and then asks, where do we go next?

What of the future? Well, *you and your dog are the future*: Spread the word by sharing your joy with your dog in public. Every twenty-five yards on every walk, stop to showcase what you've learned together — a short performance at every stop. Using handsignals will impress onlookers no end. Maybe include a few Speak/Shushes to draw attention like a Cockney barker in a street market, some rapid position changes, yo-yo Back up and Come fore, and maybe Rollover, Beg, High five, and Bow to each other, laugh, and walk on. That's all it takes for others to see what you do... and want it! Then more dogs will benefit from an education.

What do I predict for the future? I think, with all the advances in technology, we'll see an explosion of reward-only, computerized autoshaping devices to teach dogs all manner of life skills, and specifically, autoshaping will revolutionize the training of sniffer dogs. Additionally, there will be a slew of mobile phone apps that monitor a dog's behavior and location and provide ongoing, real-time guidance for people as they train their dog.

What would I like to see in the future? Oooo! I would like to see dog trainers conducting their own research on dog training, using simple test-train-test formats, to quantify the reliability and speed of their methods, to share their findings, and to foster communication and cooperation among trainers to strive to keep dog training an evolving profession.

Also, I would love to see the *gamification of dog training* in homes, parks, and classes, at games events, and in a whole new breed of dog television game shows. There's nothing more fun than having fun.

Reward-Only Electronic Autoshaping

Reward-only autoshaping machines will be the revolution of the future: to use a computer's consistency, impeccable timing, and tireless work ethic to monitor a dog's behavior and automatically reinforce desired behavior.

Way back in June 1990, I teamed up with Dr. John Watson, a child developmental psychologist, and together we developed such a machine — an AutoTrainer — which had an interactive algorithm that taught dogs to bark less. All the operator had to do was fill the food hopper with the dog's daily ration of kibble, turn it on, and press one of three buttons: "Teach Me," "Calm Me," or "Feed Me."

The machine kept track of the number of barks and mesmerized dogs with a cascade of tones that predicted eventual kibble delivery after progressively increasing periods of silence. If the dog barked, or tilted the machine, it would say, "Uh-oh," and reset the tone cascade back to the start.

Testing the machine with inveterate barkers was fascinating. All dogs stopped barking within twenty-five minutes and settled down next to the machine. John commented that the dogs were showing all the signs of "bonding" to the machine. They would lie close to the AutoTrainer and responded to the machine's soft "uh-oh" buzzes by pricking their ears and cocking their head to one side. John thought that dogs interpreted the buzzes as the machine saying, *I heard you*, or *I felt you*, as if the machine was acknowledging their existence.

Of course, we'll never know for certain, but I found it so sad to even think that, in their loneliness, the dogs were bonding to *a machine*. I always wanted to call the machine Buddy, in the sense that lonely dogs could relax at home with their e-Buddy. Indeed, one unexpected outcome was that Buddy became an expert e-therapist for relieving extreme stress and separation anxiety. Buddy's gentle, interactive algorithm broke through. It is virtually impossible for any animal to resist ongoing operant conditioning. Even dogs that wouldn't eat from chewtoys when left alone soon learned to eat kibble from Buddy's "hand." Each dog's newfound relief somehow camouflaged that Buddy was just a machine, albeit with precise training expertise.

Reward-only autoshaping has many wonderful specific applications that would make life easier and more enjoyable for dogs and their people. In addition to reducing barking, pacing, anxiety, and loneliness, an autoshaper can crack pretty much any behavior problem, such as where to eliminate or what to chew. And it can teach dogs to perform any behavior on cue, for example, Come, Sit, Down, Stay, Follow, and so on, by using a language of ultrasonic beeps, which the owner could control with a handheld remote.

The AutoTrainer became our hobby, and since my doggy passion has always been olfaction, John and I created a prototype to teach dogs odor detection. During her last year, ZouZou was my experimental dog, and within a couple of days of training, she was marking odor sources with a Sit-Stay to within twelve inches. But then she got sick, and I shelved the project. I miss her as a work colleague, a constant companion, and occasional dance partner.

My dream is to reintegrate dogs into society by autoshaping companion dogs en masse to alert to explosives and firearms. Each owner would have a GPS remote, and each autoshaped dog would wear an official vest, so they would be welcomed into stores, restaurants, pubs, and on public transport, just like the good old days. Several dogs could be stationed outside and inside schools, places of worship, railway stations, airports, sports stadiums, and concert venues. Whenever their dog alerts, all the handler needs to do is press the GPS remote and send a warning to a centralized agency, say in Omaha, Nebraska. First responders could triangulate the source from the direction of the dogs' noses. What a safer place the world would be — a world with dogs protecting us once more.

Billions of dollars have been spent on designing sniffer machines that rival a dog's nose. Researchers have come close, but the machines are heavy and huge and next to impossible to transport. Also, the machines don't have legs and can't move themselves from house to house, from room to room, and into tiny nooks and crannies.

Training sniffer dogs and their handlers is expensive and takes many months, whereas autoshaping takes only a few days and computer time is cheap. I would love for animal psychology laboratories to make autoshaping odor detection a priority, so that autoshaping sniffer dogs *en masse* becomes commonplace in the future.

Apps

I also predict loads of dog training apps — handheld, computerized dog trainers available 24/7 to teach people how to train their dog and how to praise effectively, *in real time.*

An AI app would identify *your* dog, evaluate *your dog's behavior*, and most important, *your feedback*. Plus, the app would offer *immediate and ongoing guidance*. For example, once the app hears your voice say, "Rover, Come," it would instantly check the distance between the two of you, your dog's approach speed, and your feedback, broadcasting suggested feedback for you to mimic, both words and tone — a personal "navigation system" for you and your dog's education.

Moreover, the app would keep records of overall progress, and let you know in real time whether each recall was quicker or slower than average and automatically calculate response-reliability percentages for body position changes, length of stays, heeling closeness, and speed and reliability of emergency Distance Sits.

Another application would be to monitor behavior during dog-dog play and keep a tally of the "Friendly Quotient" for both dogs to reassure owners that the dogs are enjoying play, or whether a dog is becoming stressed or amped up and it's time to settle for a while. And all while you keep a video record on your mobile.

Research on Dog Training

There are some wonderful studies on dog cognition, and there are lots of surveys on personality, temperament, and adoption statistics. Most of these studies are extremely well done and have interesting results.

On the other hand, there are many not-well-done surveys based on questionnaires given to owners regarding, for example, the effectiveness of drugs for resolving behavior and temperament problems, such as separation anxiety or aggression. These questionnaires are useful for quantifying the subjective views of respondents, but they provide little to no objective data about dogs themselves. A person's subjective impressions of their dog's behavior can be very different from objectively testing dog behavior and responses. As Dr. Beach always told me, "Look at the data." Otherwise, when results of subjective questionnaires are summarized in the media, it becomes fact, right?

Surprisingly, there are very few research studies on the speed and effectiveness of different methods of dog training, and these are so desperately needed for dog training to evolve as a science. We need fact-based studies based on quantifying behavior before and after training: *test-train-test*.

We need studies on dog trainers training dogs to objectively compare the speed and effectiveness of different training techniques, as well as to evaluate our own. But most important, we need test-train-test studies on dog owners training their dogs to objectively evaluate how easily training advice transfers from book to trainer to owner.

Consequently, years ago, under the auspices of the Association of

Professional Dog Trainers, I founded the APDT Foundation to raise money and encourage research on dog training *by dog trainers*, and to award prizes for the best *completed* studies and publish them in an online APDT journal.

Medical, psychological, and drug studies generally require thousands and thousands of test subjects, complicated experimental design with multiple control groups, and power statistics to eke out statistically significant results. By contrast, dog training research produces highly significant results because a dog's responses are observable and quantifiable, and of course, training is all about effecting *behavior change.*

As an example of how easy it is for trainers to conduct scientific studies themselves in their classrooms, *and generate highly significant results,* I once asked our SIRIUS trainers to test puppies in class at the beginning of the first week and then to administer the same test at the end of the sixth week. I presented the study at an APDT annual conference to initiate a "science track."

In the SIRIUS study, forty-seven puppies received the same test prior to and after a six-week puppy class. The test comprised response-reliability percentages for the S-D-S-St-D-St position-change sequence using handsignals only, and then verbal requests only, plus durations of Off, and Sit-, Down-, and Stand-Stay.

The various pretest scores improved by up to 1,113 percent. Moreover, when pretest and posttest scores were plotted as a histogram, they passed the most stringent of statistical tests: the "interocular trauma test." Statistician Joe Berkson used this term to describe when the results of plotted data are so obvious that they *hit you between the eyes*. That's how clear our results were. There was a massive white space between the two bell curve distributions of pretest and posttest scores.

A test-train-test format using each dog as its own control is so easy to apply without interfering with individual or class training. Video test sessions to facilitate scoring and offer proof for others. Share your results with other trainers, who can share their results with you. This is how we learn, by sharing, discussing, listening, watching, and helping one another improve. If we want dog training to be science-based, let's make science.

We would do well to question and evaluate everything in dog training. Personally, I only believe in what I see with my own eyes. Yes, I listen to other trainers' opinions and descriptions of what they do, but when trainers mention numbers, I perk up my ears shepherd fashion. And when they show me unedited video of observable, quantifiable behavior change, I am an instant believer. I encourage everyone to conduct their own studies and present them to the APDT Foundation. The research possibilities are endless.

Television

Over the years, I've written numerous treatments for dog training programs and doggy game shows for television. Several we produced ourselves on DogStarDaily.com. First up were *America's Dog Trainer* and *World's Dog Trainer*. Instead of having the same host for every season, we featured a different trainer for each episode, so that people could get advice from a variety of trainers, instead of just one. The third program was a video contest of *puppy pantomimes* — mini-plays, soaps, and short stories with most of the actors being dogs. The theme was *summer fun*.

Dog owners and their dogs are just begging for more instructive and entertaining television programs, which would revolutionize reward-based training overnight. Dynamic, exhilarating, captivating, action-packed, fun-filled, fast-moving shows could bring back sparkle, excitement, and razzle-dazzle to companion dog training. They could show puppies being trained while playing, and people teaching recalls and heeling and playing games in dog parks. I find real-time learning both fascinating and illuminating. I love watching people having fun with their dogs, and dogs having fun with their people; playing games and competitions light up the screen.

I find too many dog training shows are formulaic and focus on *problems*, with the dog presented as the "bad guy." Instead, why can't *the dogs be the stars* of these shows, which celebrate their vibrance and the intrigue of their behavior, play, and education? Dogs have an astronomical "Ahhhh! factor."

Think of all the things dogs have done for the media by tirelessly selling tacos, beer, beans, cars, and insurance in advertisements. In England, golden retriever puppies even promote human toilet paper. I mean, who else are you going to get to promote toilet paper? A bear? A bunny? A duckling? Why can't media give back to dogs and give them the opportunity to promote themselves and dog-friendly dog training?

Playing Games

Earlier I mentioned, *don't be a food-vending machine, be a slot machine.* Food and life rewards are intrinsically reinforcing, but it's your reinforcement *schedule* that magnifies their impact a thousandfold. To increase their reinforcing bang for the buck and your dog's anticipation and desire to learn, make your rewards unpredictable, variable, and bordering on magical, with the ever-present anticipation of a jackpot — the vivid memory of a buried cow's femur or discovering a "treat tree" in the park. Motivate your dog so that playing the training game with you becomes the biggest reward in the world.

Ongoing quantification during the various stages of lure-reward training is vital for offering and evaluating proof of progressive improvement. Nothing can be more fulfilling and motivating than knowing that you and your dog are succeeding. But numbers need not scare you. Playing games checks all the requisite boxes: Yes, of course, games are fun, but they also are objective quantifiers of performance, your dog's and yours. Also, playing games in class is the quickest way to teach people *how* to teach basic manners, and it significantly motivates them to practice the vital foundations of training at home, especially when they don't know the real purpose of each game.

The Praise Game

Training works much better as confidence grows and people shed their inhibitions and learn to praise like they mean it. Sometimes, getting people to thank and praise their dog is like getting blood out of a stone. Laughter is the best way I know. Laughter, belly laughter, releases serotonin and endorphins, and they both kick in quickly to replace stress and anxiety with happy feelings followed by longer relaxed periods of euphoria, focus, clarity, and calm. Acquired knowledge suddenly becomes clear, clicks into place, and allows training to zoom. Remember the seven male "dwarfs"?

Laughter is a most powerful medicine indeed. After laughing, people have neither the strength nor the desire to remain reserved, restrained, or reticent. Instead of being a person-in-public-with-a-dog-in-class, their entire attitude and demeanor changes in a second; *they become themselves and go for it.*

I always tried to get my audiences to laugh lots in seminars. In the Praise Game, I would home in on the essential issue. I would pick five judges and then tell the audience, "We're going to have a competition to see who can praise the best. I'm going to pretend I'm a Yorkie pup. It's dark and cold and raining outside, and you're in the middle of watching the World Cup, but it's your turn to 'de-pee and de-poop the pup.' It's halftime, you're eager to watch highlights, we're outside, and I'm on your side, and... *Look!* I just peed. *Praise me!!!*"

Then I would point at some unsuspecting male "volunteer." Silence. But everyone is staring at him and so eventually, "G-good... d-dog?" We would then go to the judges holding up the scores. I would make some comment, "Ahhh, we welcome the Canadian judge back today. In all honesty, you thought that was worth a three?" Then I would reassure the man. "Hey! At least I peed. Tell your wife." I used to love evaluating their "praise": "You still breathing?" "Wonderful inception. Explosive! But then *praisus interruptus.*" "Doesn't travel well." And very occasionally, "That was genuinely loving, straight from your heart, and left me with a lingering aroma of roses and kibble." Then I would invite all "my men" for a beer afterward. And did we laugh!

Games for Teaching

Like tricks, games have numerous functions, and the queen of functional games is Musical Chairs. This game perfects off-leash heeling, quick Sits, and bombproof Stays. Also dogs develop unbreachable confidence around other dogs, people acting silly, and a cheering audience, and owners naturally offer a torrent of instructive, reinforcing, and convincing feedback. What gets better than that?

Musical Chairs

Basically, owners heel their dogs off-leash, counterclockwise around a taped rectangle that has a line of alternately facing chairs down the middle. When the music stops, each owner instructs their dog to Sit (without contact) *outside* of the rectangle, and then they walk quickly to sit on a chair.

If any dog breaks their Sit-Stay, or *touches or crosses the line* of the rectangle, their owner must immediately vacate their chair, completely leave the rectangle, reinstruct their dog to Sit, and then hope that another chair has become vacated. The person without a chair may walk around the line of chairs twice, acting weirdly with silly walks, frog-jumping, or rolling on the ground (just like children), but not frightening any dog (judge's call), or getting closer than one yard. Once a chairless owner completes two laps of attempted distraction, it's sayonara — the long, slow walk to the spectator's area.

The game fosters gamesmanship, aka *common sense*. For example, heel your dog on the right so you have a direct path to the chairs, and sit the dog at least three yards back from the line so your dog cannot touch the line and so that people without a chair cannot get close to your dog. Also, vary heeling speed to create distance from the other competitors, and when only a few chairs are left, speed up if no chairs are close, but slow down if approaching a chair.

As a judge, I used to love it when a dog, sitting too close to the line, wagged their tail to cut the line: "Vacate your chair. Yes, it's fair. Read the rules." The crowd would love "tail-faults" and "paw-faults," as did I. Sounds ruthless, I know, but they help people learn *how to manage their dog in real life*. Such as in an outdoor restaurant: You want to settle your dog in a corner, or close to a wall, and looking at you, rather than close to other dogs, looking at them, and at the end of their leash.

After playing Musical Chairs half a dozen times, most owners and dogs are well prepped to waltz through a Kennel Club Canine Good Citizen Test, and they would probably get a respectable score in a KC Novice Obedience Trial.

Joe Pup Relay

The original Joe Pup Relay was a timed event. Owner and dog run off-leash from a Sit-Stay at "home plate" to first base to perform a S-D-S-St-D-St sequence, then to second base for a Sit-Stay while the owner circles their dog singing, "How much is that doggie in the window..." Then they run to third base, where the owner lies down flat on their back with arms crossed and verbally instructs their dog to lie down and says "Good dog," three times, and then rush to the finish for a Sit on home plate.

The first time I played Joe Pup was at a pet dog event near Windsor. The game was played as virtual-reality training: Each base represented *a real-life scenario*: (1) Sit at the vet's door, Down-Stay in the waiting room, and Stand for examination; (2) Sit-Stay on the tailgate of a Range Rover, while walking round the car, carrying two bags of groceries; and (3) instructing a dog to lie down while reclined on a couch.

In classes, I played many variations of Joe Pup. For example, I asked each owner to practice any behavior or trick during the week, write the name on an index card, and bring it to class. We put all the index cards in a hat, and right before each person's turn, they pulled three cards from the hat for what their dog had to do at each base. Most commands they knew, but some they had to *teach their dog on the fly*. It is so important that owners learn to just give it a go, rather than freezing or dithering. A lot of weird stuff can happen in real life, and you must act quickly and definitively.

Often owners had difficulties and would forget what to do, but the audience assistance was glorious to behold. People would offer tips, shout instructions, and urge contestants to hurry. Others would jump up and help. The success rate was outstanding because, with the clock running, people felt urgency, and with everyone in class encouraging them, they just *gave it a go.* The combination of the stopwatch and audience support inspired people *to take their best shot.* Everyone helped each other. I loved it. I've seen accountants turn into inspired athletes, causing their dogs to light up like fudged tax returns.

Games for Motivation — Not Revealing the Reveal

If a magician said, "I'm going to perform a card trick, but I won't actually put my volunteer's selected card back in the pack, I'll put it in an envelope to slip into their pocket," few people would be enamored with the "trick." (My first stage appearance was as a ten-year-old magician, and I *did* announce what I was going to do, but it never happened; the egg always broke, and I always picked the wrong card, and so it turned out to be my first comedy routine. My mother was mortified by the laughter, but I became hooked on audience feedback.)

My favorite games are ones where the players have no idea what's going to happen or even what they're learning.

Doggy Dash

In a doggy dash, two dogs race side by side, and the first one to sit *completely across the finish line* wins and goes through to the next round. During the quarterfinals of a K9 Games' Doggy Dash (with seventy-two dogs in the first round), in front of thirty thousand people in the Toronto Sky Dome, I remember when a basset beat a border collie. The border collie was across the finish in a flash and then ran circles round their owner, barking and jumping. A good five seconds later the basset lumbered across the finish and sooooo measuredly … sat, and so, advanced to the quarterfinals.

In the doggy dash, successful recalls are a given because the dog starts in a Sit looking at their person, and the owner's exuberance is off the charts. The real purpose of the game though, is *to motivate teaching quick Sits.* Sometimes I find it exhausting, repeatedly telling people about the importance of a reliable, lightning-fast sit. However, just one race won, but then lost, because of a slow or no sit, and it won't happen again. People practice.

Additionally, as with all games, doggy dash *teaches* people *to pay attention* to my instructions (the rules of the game). In one race, one dog came and sat, but with their tail *on* the finish line and so the slower dog won. Again, the audience went wild. A tail-fault! But again, it teaches people to be more mindful of where they sit their dogs, for example, not close to an open door or the edge of a sidewalk.

Doggy Dash Variations

I invented loads of doggy-dash variations for classes and workshops. Any type of timed game or race catalyzes the best praise from people and joyful participation.

In a slow dash, both dogs must cross the start line within three seconds and maintain forward movement, but it is the last dog to cross the finish that wins. Dogs are disqualified if they stop for a palpable second or leave their "racing" lane. This race is nail-biting to watch and extremely difficult to master, but afterward, with their newfound precise control over their dog's speed, *forging and pulling on-leash become history.* Also, there are situations that require a slow recall; for example, when a family member lets your dog join a cheese-and-wine party, most people are drinking red wine, and you have wall-to-wall beige carpets.

In creep-and-limbo, two dogs race by having to GI Joe down the course and crawl *under* three low jumps. In the backward send-out, the owner remains behind the start line and directs their dog to run backward down the straight and across the finish line. The biggest problem is a dog careening out of their racing lane and bumping into the other dog. If that happens, the owner is allowed to continue after calling their dog back over the start line to begin again. A rollover race would always bring tears to my eyes; once, a Newfie rolled out of her racing lane (disqualification) and over a border terrier. She won't do that again; the disturbance under her belly was disconcerting!

The crazier the games in puppy classes, the easier the *puppies take craziness in stride during adulthood* and learn to *listen to their owners.*

Beer Heeling and Kibble on a Spoon

John Rogerson, a close friend and an extraordinarily talented dog trainer, developed a hilarious leash-control game — *beer heeling*. Multiple contestants raced a course on his local's lawn, but with a pint of beer in their leash hand. (*Local* means "pub.") If they spilled more than one dimple of beer from their dimpled stein beer mug, they had to finish the beer and then race again. When visiting John, I perfected this game; the secret to success is a Groucho Marx hurried-but-no-spill walk. Never in my life have I seen *leash-walking improve so much in such a short time*. Dogs became glued next to their owner's side. Why? Because everyone paid *riveted attention* to any tension in the leash and gave a steady stream of *urgent and instructive feedback* to keep their dog in heeling position.

We played a kibble-on-a-spoon variation of this the very first time I held the K9 Games (no beer allowed at Fort Mason). Eight contestants, each with three pieces of kibble on a spoon, raced a course with various distractions, such as going through narrow gates, hot dog alley (between two rows of hot dogs swinging on strings), and a short pit stop (a ten-second Sit-Stay in a gated ex-pen with three pitties in Down-Stays). Using a combination of extreme skill, common sense, and forethought, Suzi Bluford and Streaker won this by a mile and were the only ones to qualify by not spilling a single piece of kibble. At the start, she kept Streaker in a Sit-Stay and watched as the others raced off and bunched up and bumped each other at the first gate, causing everyone's kibble and all the dogs to go flying in all directions, and then she set off and slowly walked the course with a clean path. It was such a lovely event; I still watch the videos from time to time.

Biscuit Balance

At the end of week three of Puppy 1 classes, I would hold up a ribbon and announce, "Next week, the puppy that can balance a biscuit on their nose for the longest time wins this red ribbon." (Red? I'm British.)

One little boy told me he practiced with Jack, the family dog, all evening, but then his dad sent him to bed early, saying that Jack was tired and had to rest up for puppy classes. However, the son could hear his dad working with Jack until the wee hours. During class, the record climbed to over a minute, and when Jack's turn came, I said, "Who's going to work with Jack?" I looked toward the little boy, but his father jumped up and said, "I will. I'm much better." I thought, *No. You're. Not!*

After the father commanded Jack, "SIT!! Staaaaaaaay," and gingerly balanced a biscuit on Jack's nose and then slowwwly moved his hands away, I started my stopwatch and then tossed a treat to the side. Jack explosively turned his head and the biscuit fell to the floor. "Not bad," I said. "That's 2.3 seconds." After everyone else had had a go, I asked the little boy if he wanted to try. "Come on, I'm sure you can beat your dad." And he did. Jack balanced the biscuit for two minutes and thirty-seven seconds. The little boy was so happy, and to his credit, his dad congratulated his son and gave him three high fives.

I then asked the boy if he could read the gold lettering on the ribbon. He squinted his eyes and read, "SIRIUS Puppy Training... First Place... Longest... Sit-Stay." Can you imagine if I had said, "Homework tonight is Sit-Stays"? No one would have practiced, and the dogs would have been all over the place.

Simple games are exceptional for motivating owners to practice. It's the best way I know to get competition-ready, rapid, and reliable responses with position changes, solid stays, and repetitive one-step Heel-Sits and Come-Sits — *the base building blocks of all training.* Owners will diligently practice most anything with ribbons and rosettes as potential prizes. I bet a lot of ribbons, rosettes, and diplomas are still on fridges or mantelpieces long after the dogs have passed away. I know mine are. Claude's participation "diplomas" and Hugo's rosettes are still hanging in my office.

Games for Group Camaraderie

To foster camaraderie and friendly competition, practically any game can be played as a team game, as a family, or as one large group.

Relays

Even a simple recall relay is exhilarating to watch, with every team member *cheering on their fellow teammates* and the spectators going wild: Two teams compete head-on, four dogs and owners to a team, calling their dogs in succession, and when all dogs are in Sit-Stays across the finish line, all dogs are instructed to lie down.

When people mastered beer heeling without spillage, John turned it into a relay with four teams racing at the same time and four dogs and owners to a team. The changeovers demanded beyond *exquisite control of both dogs.*

Joe Pup is an extremely useful game to play as a relay. Two teams race at the same time with four people and one dog per team. The dog runs from one person to the next, performing a short routine with each. The various routines are not revealed until 9 a.m. on the morning of the event. This game allows owners to practice *teaching family members and friends* how to work their trained dog. Just as when teaching Go to with children, once another person is controlling the dog, the owner must be very careful not to move or talk, otherwise their dog might turn to look and run to them.

The woof relay is fast and dramatic. Five dogs and owners to a team and each has to instruct their dog to woof three times *and then shush* in succession. In Japan, all our K9 Games events were on national television, and some were the most-watched programs. I remember one winning woof relay team being interviewed: "But doesn't this make your dogs bark more in your apartment?" The captain instructed her team to Woof and Shush their dogs several times in succession and said, "If they bark, we ask them to Shush." The interviewer then asked, "How did you teach that?" Her reply: *"Ichi, ni, san, yon!"* — 1-2-3-4. Then she demonstrated how to teach Speak and Shush to an audience of millions! I loved it.

Families

It's wonderful to train a dog as a family, as families play board games together. In fact, years ago, Terry Ryan and I created a dog training board game — *My Dog Can Do That!* These days, I bet you can't buy a copy for less than fifty bucks on eBay.

My favorite family game by far is round-robin go-tos because the game naturally transmogrifies into many other games, all of which offer free recalls and strengthen the psychological bungee cord between dogs and people, as well as fostering a spirit of cooperation and togetherness.

Once your dog has learned the names of family members and friends and will go to them before being called, walks become even more enjoyable and *a better physical workout for dogs.* Family and friends can spread out on a two-mile trail walk and send the dog back and forth between each other, so that the dog runs up to five or ten miles and returns home physically satiated.

Families can practice "search-and-rescue" games: "Sam, Find Michael," and after each find, everyone laughs and giggles as Michael hugs Sam, and Sam rolls over for a chest massage. Aside from hide-and-seek being a wonderful indoor game on rainy days, being able to communicate to your dog to find a person, another dog or animal, or an object has many uses. The task itself is simple for a dog's nose, if only they know *what* to search for: "Sam, Find Gramp." "Sam, Find my glasses." "Sam, Find the car." Currently, Samuel Squirrel is learning, "Sam, Find Mousie" — the kitty. Next up should definitely be, "Sam, Find my passport!"

Leader's Choice

A wonderful group game is leader's choice, which we would often take outside to the park. Owners, with dogs on-leash and sitting at heel, stand in a long line spaced by arm's length. The leader says, "Forward," and everyone steps forward at the same time and walks in step. When the leader says, "Halt," everyone stops, and the leader then performs a very short, cued routine, such as Back up, Playbow, figure eight in motion, a Rollover, or a Woof. Then everyone in the line follows suit.

The leader then walks to the back of the line, whereupon the previous second-in-line becomes the leader and instructs a *different* routine.

This is a wonderful *social event* with *people helping each other.* Training on the trot in a group gives people enormous confidence.

To end this section, I would like to give a shout-out to Roy Hunter, the leader of the 1960s London Dog Trainers "Rat Pack," along with John Uncle, Alan Menzies, and honorary members John Rogerson (from North of Luton) and Terry Ryan (US). One fun-loving crew; always playing practical jokes on each other, and all playing games with dogs in their workshops and classes. Both Roy and Terry wrote a variety of wonderful little books on fun and games with dogs.

Roy was modest, unassuming, and utterly unpretentious. He had a heart of gold, helped everyone, and did enormous things for dogs in his career. Roy oversaw the London Metropolitan Police Force dog section and developed agility and nose work as part of their training. Some years *before* agility was "first" demonstrated at Crufts in 1978, Roy had demonstrated agility at the same venue with homemade equipment (largely oil drums) and his team of police dogs and their handlers.

In honor of Roy and his good people, it's high time we resuscitated laughter and fun and games in dog training.

Gamify Dog Training

In addition to reinforcing your dog like a slot machine, teaching lots of tricks, and playing lots of games, you can go one step further by imagining training as a video game that you play with your dog — *gamifying the entire process of learning to train.*

The interactive algorithms in video games are the pinnacle of reinforcement. They can steal people's attention and hold their brains captive for hours on end. One massive reward is allowing players access to play at progressively higher levels based on their mastery of play. There are "treasure boxes" along the way that reveal secrets of the game and enable the player to "power up" and play with more powerful tools, making it easier to navigate higher levels. Eventually, players become hopelessly addicted and keep trying to surpass their best scores or the scores of others.

Basically, this is how I tried to write this book. Each chapter contains a treasure box or two that reveal the secrets of the method so you can "power up" and reach higher levels of training. You can use more powerful communication and reinforcement tools that make it easier and quicker for you and your dog to develop a mastery of training. As you reinforce your dog's responses, and your dog's masterful responses reinforce you, eventually, playing the "training game" together gets its hooks into both of you.

Here are the ten levels of the training game:

Level 1: Focuses on *early enrichment and socialization*, starting in the breeding kennel and continuing at home as you build your dog's *confidence* around unfamiliar people, dogs, and situations in the world of humans. This provides your dog with a safety shield to *prevent* adolescent-onset fears and phobias and lack of confidence.

Level 2: Comprises the time-consuming and difficult transition from training on-leash to training off-leash — but *you* get to skip level 2 because you *train your dog off-leash from the outset.*

Level 3: Contains an enormous treasure chest — *teaching ESL and testing comprehension* — so that you may truly talk to your dog *and your dog understands*. You can now communicate with your dog on a different level, which is exactly where you go....

Level 4: Enables you to *power up food rewards* by decreasing their frequency and making rewards unpredictable, while always preferentially reinforcing better-quality responses.

Level 5: Offers treasure box after treasure box, each one containing yet another *life reward*, especially the golden rewards — *putting behavior "problems" on cue.*

Level 6: Reveals the long-sought secret of how to use your *words* to eliminate the hordes of bothersome behavior, training, and temperament problems that prevent access to the next levels, where unruly, hyperactive, and noisy canine companions would not be welcomed.

Level 7: Holds the sanctuary for *on-demand extreme reliability* — teaching a dog their *formal name* and using persistent insistence to teach *learned relevance*.

Level 8: Bestows upon you the most effective training tool in the realm — *emergency distance commands*, which enable you to use *distance rewards*: "Fido, Sit. Good lad. *Free dog!*"

Hidden Trap: Beware! If you aren't careful, you and your dog could wander down a wide and windy rabbit hole leading to an enormous room crammed with physical and aversive tools. Knowing that this is a dead end, you help your dog escape by staying calm and using words for instruction: "Fido, Sit. Fido, Come. Rover, let's go."

Level 9: Presents the *Game Show*: a plethora of tricks, games, and friendly competitions to gamify training itself. This level is always full of happy dogs and owners, and so it may feel like the end of your journey, but you know there's one more level, and it's the hardest to reach....

Level 10: Try as you might, reaching level 10 requires something you can't get from your dog, not directly, nor from any treasure chest or training tool. This elusive attribute is confidence and belief in yourself: *The Force in Dog Training*. Achieve that, and you and your dog will win the game.

The Force in Dog Training

Even after successfully learning how to use lure-reward training, many owners still don't have full *confidence* or *belief* in their own abilities, especially in certain real-life situations. Most dogs can recognize that. When owners get to this point, I try to help them develop invincible confidence by teaching them that they have *the Force*. In these exercises, I put owners in impossible situations and work with them until they have conquered all. One of these I call the *hot dog recall*. The owner has to remain seated in a chair while a child seduces their dog away by waggling a hot dog. The owner has to call their dog to

Come-Sit and take hold of their collar, while the rest of the class looks on. This is a not-uncommon scenario for park picnics, and owners must learn how to deal with it.

At first, most owners don't utter a sound. Then they offer a series of interrogatives, "Rover, Come? Roverrrrr, Come? Rover, Come ... here?" Hardly any owner ever thinks of switching to their dog's formal name, even though we've spent weeks practicing. Silence again and maybe a meek query, "What am I meant to do?"

"Get your dog to come and sit," I say.

The owner will call tentatively a few more times, and then something happens.... Some owners raise their voice, others lower it, but after taking a deep breath, their next instruction is different — literally turgid with insistence — and their dog quickly glances back at them.

Instantaneously, I "grab" that glance and verbally and mentally help tow the dog toward us, like reeling in a blue fin tuna: "Gooood dog! Good dog! Pup, pup, pup, pup, pup, come on. Come on. That's the way!" Clapping my hands, I tell the person, "Call your dog! Call him! Woooohoo. Yes! Goooood boy. Tell him Sit. Sit! Handsignal! Signal! Collar! You've got him. Well done! *Hot* dog!"

As the owner succeeds at getting their dog to sit and holding their collar, I quickly also take hold of the dog's collar to make sure the owner doesn't let go. Then the child runs up to give the dog part of the hot dog. Then I say, "Incredible! Let's do it again."

By the time, bit by bit, the hot dog has disappeared, most owners can do it on their own. Whereupon, I stand in front of them and, after asking permission, place my palm on the top of their head and intone in a quiet voice, "You have *The Force in Dog Training*. Don't you ever forget. Well done!"

At which point, the entire class erupts in an explosion of laughter, cheering, and applause. And we have graduation. This is level 10!

A Final Wish

You know what I would really love? Sometime in the future, while wandering along the beach in Barwon Heads, or ambling down the

Coast Highway in Encinitas, or sitting outside the Famous Cock Tavern (across the road from the Hen and Chickens), or even wandering down Main Street USA, let's say in Iowa City... I would love to turn a corner and see some thirty people and their dogs, all wearing the same pink T-shirts and bandannas, playing leader's choice.

Or maybe *you*, striding down a sidewalk, with your spritely, seventeen-year-old dog heeling magnificently and looking up at you with pride and affection, and then suddenly, you abruptly stop, your dog moves round to sit-front, you open your arms, your dog stands on their hind legs, and the two of you dance a tango, with abrupt head-turns and all... and then, as quickly as it started, it stops. You both bow to each other and then heel away. That would make me feel so happy. It would make my day, my week, my life.

I'm around a lot of dogs in training. And after training, the dogs leave to do the jobs they were trained for, and I miss them. The next dog I'll miss is Taffy; I am convinced that ZouZou's soul lives in her heart. I never think of training a dog as a chore to be rushed but rather a delight to be savored, one that I wish would last forever. I think of training a dog as no different from teaching a child to read, write, play chess, build Lego, kick a football, ski, and to greet and chat to people in a way that makes *them* feel just a little better about *their* day. I wish that, too, would last forever. It's what life's all about. It is life. Especially when the child internalizes much of your advice and guidance, and surpasses your skills at chess, skiing, and *offering advice*.

Thank you, Jamie.

Resources

Some of the training procedures I've described in this book can be difficult to visualize, and so I've created a free video course at DunbarAcademy.com, exclusively for readers of *Barking Up the Right Tree*. To access this course, go to https://dunbar.info/RightTree or scan the QR code below.

Dunbar Academy also has hundreds of hours of dog behavior and training videos.

Puppy Training

Arden, Andrea. *Dog-Friendly Dog Training*. New York: Howell Book House, 2007.

Dunbar, Ian. *Before & After Getting Your Puppy*. Novato, CA: New World Library, 2004.

Dunbar, Ian. *Dr. Dunbar's Good Little Dog Book*. Oakland, CA: James & Kenneth Publishers, 2003.

Dunbar, Ian. *How to Teach a New Dog Old Tricks*. Oakland, CA: James & Kenneth Publishers, 1991.

Miller, Pat. *Positive Perspectives*. Wenatchee, WA: Dogwise Publishing, 2006.

Dog Tricks

Kay, Larry, and Chris Perondi. *The Big Book of Tricks for the Best Dog Ever*. New York: Workman Publishing, 2019.

Sundance, Kyra. *101 Dog Tricks*. Beverly, MA: Quarry Books, 2007.

Sundance, Kyra. *101 Dog Tricks: Kids Edition*. Beverly, MA: Quarry Books, 2014.

Dog Behavior and Training

Clothier, Suzanne. *Bones Would Rain from the Sky*. New York: Grand Central Publishing, 2005.

Donaldson, Jean. *The Culture Clash*. Wenatchee, WA: Dogwise Publishing, 1996/2013.

Horowitz, Alexandra. *Inside of a Dog*. New York: Scribner, 2010.

McConnell, Patricia. *The Other End of the Leash*. New York: Ballantine Books, 2003.

Index

About the Author

Dr. Ian Dunbar is a veterinarian, animal behaviorist, dog trainer, and writer. He received his veterinary degree and a special honors degree in physiology and biochemistry from the Royal Veterinary College at the University of London and a doctorate in animal behavior from the psychology department at the University of California at Berkeley, where he spent ten years researching olfactory communication, the development of hierarchical social behavior, and aggression in domestic dogs. Dr. Dunbar is a member of the Royal College of Veterinary Surgeons, the California Veterinary Medical Association, the Sierra Veterinary Medical Association, and the Association of Pet Dog Trainers (APDT), which he founded in 1993.

For over seven years, Dr. Dunbar wrote the American Kennel Club *Gazette*'s behavior column, which was voted Best Dog Column for a number of years in succession by the Dog Writers Association of America. He has written numerous books, including *Before and After Getting Your Puppy*, *How to Teach a New Dog Old Tricks*, and *Dr. Dunbar's Good Little Dog Book.*

Dr. Dunbar has been lecturing to veterinarians, dog trainers, dog professionals, and dog owners since the 1970s. In 1982, Dunbar founded SIRIUS® Puppy Training, which featured the first off-leash puppy socialization and training classes in the world.

Dr. Dunbar's techniques completely changed the way dogs are trained. His unique lure/reward, off-leash training techniques provided a delightful alternative to on-leash training. There are very few dog trainers who have not been strongly influenced by Dr. Dunbar's fun-and-games, from-the-animal's-point-of-view, dog-friendly dog training. He lives in Berkeley, California.

DunbarAcademy.com • DogStarDaily.com • SiriusPup.com

NEW WORLD LIBRARY is dedicated to publishing books and other media that inspire and challenge us to improve the quality of our lives and the world.

We are a socially and environmentally aware company. We recognize that we have an ethical responsibility to our readers, our authors, our staff members, and our planet.

We serve our readers by creating the finest publications possible on personal growth, creativity, spirituality, wellness, and other areas of emerging importance. We serve our authors by working with them to produce and promote quality books that reach a wide audience. We serve New World Library employees with generous benefits, significant profit sharing, and constant encouragement to pursue their most expansive dreams.

Whenever possible, we print our books with soy-based ink on 100 percent postconsumer-waste recycled paper. We power our offices with solar energy and contribute to nonprofit organizations working to make the world a better place for us all.

Our products are available wherever books are sold.

customerservice@NewWorldLibrary.com
Phone: 415-884-2100 or 800-972-6657
Orders: Ext. 110 • Catalog requests: Ext. 110
Fax: 415-884-2199
NewWorldLibrary.com

Scan below to access our newsletter
and learn more about our books and authors.